# Pyrotechnics in Industry

# Pyrotechnics in Industry

Richard T. Barbour
Pyrotechnics Design Engineer
Space Shuttle Program

McGraw-Hill Book Company
New York St. Louis San Francisco Auckland
Bogotá Hamburg Johannesburg London
Madrid Mexico Montreal New Delhi
Panama Paris São Paulo Singapore
Sydney Tokyo Toronto

**Library of Congress Cataloging in Publication Data**

Barbour, Richard T

Pyrotechnics in industry.

Includes index.

1.Explosives. I.Title.

TP270.B29 662'.2 80-11152

ISBN 0-07-003653-5



1234567890 KPKP 8987654321

The editors for this book were Jeremy Robinson and Ann Gray, the designer was Mark E. Safran, and the production supervisor was Teresa F. Leaden. It was set in Baskerville by J. M. Post Graphics, Corp.

It was printed and bound by The Kingsport Press.

# Contents

**Preface** vii

**1 Pyrotechnics—Who Needs 'Em?** **1**

**2 Pyrotechnic Materials** **5**

Explosion and Detonation 5
Explosive Classifications 6
High or Secondary Explosives 6
Deflagrating Propellants 8
Priming Materials 8
Time-Delay Compositions 8
Liquid Explosives 9

**3 Initiators, Detonators, and Primers** **11**

Electric Initiators 11
Explosive-Energized Initiators 15
Ballistic Hot-Gas-Energized Initiator 18
Mechanical-Energized Initiators 19
Laser-Energized Initiator 21
Detonators 22
Primers 23

**4 Linear Pyrotechnics and Severance Systems** **27**

Mild Detonating Cord (MDC) 29
Shielded Mild Detonating Cord (SMDC) 34
Confined Detonating Cord (CDC) 39
Detonating Cord 40

**5 Shaped Charges** **41**

Monroe Effect 41

Conical Shaped Charge (CSC) 47
Linear Shaped Charge (LSC) 53
Flexible Linear Shaped Charge (FLSC) 59

**6 Cartridge-Actuated Devices 75**

Separation/Release Devices 76
Powder-Actuated Fastening Systems 87
Cartridge Starting Systems for Diesel Engines 94
Thrusters, Retractors, and Pin-Pullers 103
Switches and Valves 109
Electric Utility Products 113

**7 Specialized Pyrotechnic Devices and Systems 129**

Safe-and-Arm Devices 129
Pyrotechnic Photoflash Bulb 134
Crew Escape Systems 137
Friction Initiated Devices 142
Hot Patches 148
Inflation Systems 150
Oxygen Generator (Oxygen Candle) 166

**8 Quality Assurance and Control 175**

Development Testing 175
Qualification Testing 176
Lot Acceptance Testing 178
Preflight Verification Test (PVT) 182
Mandatory Inspection Points (MIPS) 182

**Index 183**

# Preface

A recent survey by an engineering trade journal revealed that more than sixty percent of the engineers in the United States are working on programs that utilize pyrotechnics. Most of the pyrotechnic knowledge acquired by these engineers was gleaned from on-the-job training. Colleges have no classes on the subject, and only a few highly specialized libraries have one or two reference volumes that are so technical in content the reader must already be familiar with pyrotechnology to comprehend them. There is no reference material available to the layman. The author's purpose in writing this book is to fill the pyrotechnic reference material void that is sandwiched between no material at all and a limited amount of highly specialized technical jargon.

Employing the unassuming language of an encyclopedia, the author discusses the major types of explosives, propellants, powders, etc. with emphasis on their relative performance characteristics and how these characteristics are utilized to perform useful work, rather than dwelling on their chemical compositions and associated reactions. The text is supplemented with simplified illustrations presenting the rudiments of the many pyrotechnic principles. After each principle has been thoroughly explained, the discussion incrementally advances to its present day state-of-the-art. This approach has proven time and again to hold the reader's interest instead of frustrating him with a vernacular completely foreign to him.

With the ever-expanding industrial use of pyrotechnics, it is only a matter of time until engineering colleges and universities begin offering curriculum leading to a baccalaureate degree of Bachelor of Science in Pyrotechnic Engineering.

It is the author's wish that this book will help accelerate public awareness and understanding of pyrotechnics as a technology working for them every day rather than just entertaining them with midsummer fireworks spectaculars.

The author gratefully acknowledges the following individuals and companies for their assistance in transforming the idea of *Pyrotechnics in Industry* into reality:

Brian K. Hamilton, David W. Murphy, Allstate Insurance Company, AMP Incorporated, Boeing Commercial Airplane Company, Controlled Demolition, Inc., Desa Industries, Diamond International Corporation *(Diamond Match Division)*, H. B. Egan Manufacturing Company, Explosive Technology *(a subsidiary of OEA)*, GTE Sylvania *(Photoflash Plant)*, Hi-Shear Corporation, Jet Research Center, Olin Corporation *(Signal Products Division)*, Pyronetics Devices, Inc. *(a subsidiary of OEA)*, Rocket Research Company, Rockwell International *(Space Systems Group)*, Scott/A-T-O Inc., Space Ordnance Systems *(a division of TransTechnology Corporation)*, Talley Industries of Arizona, Inc., Westinghouse *(Power Circuit Breaker Division)*.

**R.T.B.**

# Pyrotechnics in Industry

chapter 1

# "Pyrotechnics– Who Needs 'Em?"

Since the close of World War II, few industries have experienced as much growth as that which utilizes the precision application of pyrotechnic energy to perform useful work. Yet, except for the small number of people who have played an active role in this growth phenomenon, few others even know it exists. Unintentionally, details of this new technology have been one of industry's best-kept secrets. Institutions of learning have been unmindful of disseminating this new technology to either their undergraduate or post-graduate engineering students.

The diverse application and efficiency of pyrotechnics are widespread. Perhaps this is because many of its applications are associated with emergency situations or are backup to more conventional systems, e.g., electrical, hydraulic, pneumatic. For example, on the sides of all military aircraft and helicopters we have all seen the familiar yellow arrow with the black border and the word RESCUE painted within in black letters. The arrow usually points to a handle or door behind which is mounted a handle. Few of us have ever seen what happens when one of those handles is pulled. In such an emergency rescue situation few people ever notice the pyrotechnic initiation handle being pulled, followed a few milliseconds later by a hatch, window, or door being efficiently jettisoned a safe distance away from the rescue area. During the aftermath questioning, few witnesses will recall the entire sequence of events. The fact that they don't is silent tribute attesting to the efficiency and reliability of a pyrotechnic rescue system. Conversely, if the initiation handle was pulled and the emergency access panel failed to be jettisoned, one can rest assured that that particular detail would be the main recollection of every witness to the emergency.

A diesel truck stops for some minor repairs that will necessitate engine shutdown for some time. After the repairs are completed, the driver discovers there is insufficient air pressure in the primary starting system to crank the engine. Fortunately for this driver, the diesel truck is equipped with a backup pyrotechnic cartridge starting system. A cartridge is inserted into a breech at the rear of the engine and is then initiated by pulling a knob in the cab. The cartridge-generated, high-pressure gas is automatically manifolded into the basic compressed-air starting system. Forerunners of this system constituted one of the first precision applications of pyrotechnic energy performing useful work. This occurred during the 1930s when cartridge starters replaced the hand-cranked inertia flywheel starting systems on aircraft engines. Although most reciprocating aircraft engines today utilize onboard electric starters, some jet engines still employ the pyrotechnic cartridge starting systems.

Astronauts place a seismic recorder on the lunar surface to record data for subsequent earth processing. Scientists interpreting the data will be able to ascertain the makeup of the local crust of the lunar surface. Not being sure a moonquake will occur during their short stay, the astronauts have brought with them a pyrotechnic thumping device that, when placed on the lunar surface and initiated, will generate moonquake-type vibrations that are picked up by the seismic recorder. This technique was perfected on earth utilizing more sophisticated recorders and much higher intensity pyrotechnic thumpers to assist geologists prospecting for petroleum and natural gas.

The now illegal use of dynamite, or other explosives, to kill fish spawned the idea of using depth charges to sink enemy submarines. In doing so, some interesting observations were made of the way heavy steel plates, which ordinarily would crack or tear if bent to such extremes, endured severe deformation and yet did neither crack nor tear. From these observations, and further experimentation, it was learned that the incompressibility of water made it an ideal medium for transferring the shock waves from the explosive to the steel plates. These tests provided the fundamental technology for the explosive-forming industries whose products have widespread application, including nuclear reactor and aerospace components.

The word "pyrotechnics" is literally defined as the art of making fireworks; however, since the close of World War II, pyrotechnics refers to a broad family of sophisticated devices and systems utilizing explosive, propellant, and/or pyrotechnic compositions. Specifically not included are bombs, warheads, land mines, and other munitions, which are commonly associated with the military term "ordnance."

Pyrotechnics offer a self-contained energy source that possesses perhaps the highest work potential (exclusive of nuclear energy) in the smallest volume and with minimum weight. In addition, pyrotechnics provide instantaneous operation upon initiation; their performance characteristics can be closely controlled; they are highly reliable and safe to handle when compared to other types of storable fuels (i.e., petrochemicals); and they possess the added asset of long-term storage capability. With these fine attributes, one can readily understand why the automotive industry has turned to pyrotechnics as the energy source to initiate and inflate their crash-safety air cushions.

The total number of pyrotechnic devices used per Apollo varied with each mission, but each one had at least 310 as compared to 139 for each Gemini and only 46 per Mercury to perform a myriad of onboard, inflight, timed, and controlled tasks automatically or on command in each of the spacecraft's systems. No failure of any pyrotechnic device was detected during any of the Apollo missions. All devices required high reliability, confidence, and safety. Most were classified as either crew safety critical or mission critical, because improper operation or failure to operate could have resulted in loss of crew, in failure to meet a mission objective, or in an aborted mission. Confidence in the Apollo pyrotechnic devices and systems was further enhanced with maximum use of redundancy. When complete device or system redundancy was not possible because of space or weight limitations, redundant cartridges in the same device or single cartridges with dual initiators (see Chapters 2 and 6) were used. Two separate and electrically independent systems operated in parallel and provided complete redundancy in the firing circuitry.

The next few chapters discuss the major pyrotechnic principles, material characteristics, and initiation concepts that constitute the "building blocks" of pyrotechnic devices and systems. Altogether, the final chapters might well have been entitled "Pyro-Techniques," for it is here that their real capabilities and benefits are presented for greater public understanding and appreciation of this new technology that is dedicated to serving mankind through the precision application of pyrotechnic energy via pyrotechnology.

By and large, the aerospace industry has been the forerunner in the diversification of pyrotechnology. It has been the leader in committing huge quantities of money to pyrotechnic research and development programs. The government, through these same aerospace companies, has also invested large sums of money in basic pyrotechnic research. Most commercial applications of pyrotechnology had their origins in these projects. This fact will assist the reader in understanding why so many of the examples throughout this book are from the aerospace industry.

chapter 2

# Pyrotechnic Materials

The first recorded pyrotechnic material was used in war and was called "Greek fire," invented by Kallenikos in A.D. 673. Although not strictly an explosive, it was used with devastating success in several battles in the Mediterranean, and was a primitive but highly effective incendiary material probably based on petroleum oils and sulfur. Its secret was lost with the fall of the Byzantine Empire in A.D. 1453.

The origins of gunpowder (or black powder) are obscure, but the Chinese were probably aware of the properties of saltpeter (potassium nitrate) as early as the Chin Dynasty (221–207 B.C.), although they did not develop its use beyond the fireworks stage until about the twelfth century. In the West this material was known as "Chinese snow," and the first recorded Western experimenter to establish the formula for gunpowder was an English monk, Roger Bacon. In A.D. 1245 he recorded it in his *De Secretis Operibus Artis et Naturae (Secret Works of Art and Nature)*.

Its first use as a gun propellant followed in about 1320, and English troops used cannon against the French at Crécy in 1346. The formula has changed somewhat since then, and today the proportions used are 75 percent saltpeter, 15 percent charcoal (carbon), and 10 percent sulfur. It is not now used as a gun propellant as it burns too quickly (about 1,312 feet [400 meters] per second), its residue fouls the bore of a gun, and it produces too much smoke. On the other hand, its rate of combustion is too slow to produce the shattering effect of a high (or secondary) explosive. It is, however, widely used in fireworks and blank cartridges, and it was used by the Russians in the retro-rockets of their planetary surface probes sent to Mars.

## EXPLOSION AND DETONATION

When an explosive substance is set off, it undergoes decomposition and releases large quantities of gas and heat. Explosion is a fast combustion, the burning spreading layer by layer through the material at the comparatively slow velocity of up to 1,312 feet (400 meters) per second, and although its rate increases with increasing pressure, it can be controlled. This reaction is often called "deflagration."

In a detonation reaction there is extremely rapid burning which produces a supersonic shock, or detonating wave, in the explosive substance. The detonation velocity is a characteristic of the explosive material itself and is unchanged by changes in pressure. It is usually between 6,500 and 29,500 feet (2,000 and 9,000 meters) per second. The detonation wave produces a very high pressure, about 1,300,000 pounds per square inch (91,390 kilograms per square centimeter), which exerts a severe shattering effect on anything in its path. The gases formed travel in the same direction as the detonation wave, so a low-pressure region is created behind it.

Explosives which react by deflagrating are called "low explosives" or "propellants"; they generate huge volumes of gases during decomposition at a relatively slow enough rate that they are extensively used in cartridge-actuated devices (see Chapter 6).

Explosives that react by detonation have the ability to shatter and are called "high explosives" or "secondary explosives." They are used extensively in linear explosives (see Chapter 4) and shaped charges (see Chapter 5).

## EXPLOSIVE CLASSIFICATIONS

Explosives can be classified according to their properties and uses and are defined in seven classes as follows:

| Class | Classification |
|---|---|
| 1 | Gunpowder |
| 2 | Nitrate mixtures |
| 3 | Nitro-compounds |
| 4 | Chlorate mixtures |
| 5 | Fulminates |
| 6 | Ammunition |
| 7 | Fireworks |

There are other important forms of classification to define transporting, storage, and fire-fighting hazards.

## HIGH OR SECONDARY EXPLOSIVES

The essential properties of explosives are the velocity of burning or detonation, the explosion temperature, the sensitivity, and the power. For the first three classes above, an absolute measurement is possible, but for the others it is usual to compare the explosive with a standard such as picric acid. Picric acid is taken to have a "value" of 100, with more sensitive or less powerful explosives having values lower than 100. The more sensitive (primary) explosives with values of around 20 are used as initiators. Their impulses can set off the intermediary (those of moderate sensitivity, about 60), which in turn will initiate the reaction in the main charges (80 or above, i.e., secondary or high explosives), which are the least sensitive. An in-depth dissertation into the chemical reactions of explosives is beyond the scope of this book; however, a short discussion of the more important variable characteristics of a representative group of six typical high explosives, including their relative power and sensitivity ratings, will benefit the reader in understanding how these various characteristics are put to practical use in subsequent chapters.

*Nitroglycerin* is a highly dangerous liquid explosive that was first prepared in 1846 by Sobrero in Turin, Italy, by nitrating glycerol with a mixture of nitric and sulfuric acids. In 1865 Alfred Nobel found that the liquid could be used safely if it was first absorbed in kieselguhr, a form of diatomaceous earth formed by the fossil remains of single-celled algae. He also succeeded in solidifying it by adding 8 percent nitrocellulose to form a gel. Nitroglycerin detonates at 25,426 feet (7,750 meters) per second, has a power rating of 160, a sensitivity of 13, and an explosion temperature of 8,000°F (4,427°C).

TNT, *trinitrotoluene,* is made by reacting toluene with a nitrating mixture of nitric and sulfuric acids. It has a power rating of 95 and a sensitivity of 110, and its velocity of detonation is 22,965 feet (7,000 meters) per second. Although it has little use in the application of precision pyrotechnology, it is included in the group mainly because it is one of the better-known explosives.

PETN, *pentaerythritol tetranitrate,* is produced by treating pentaerythritol first in nitric acid and then in acetone. It is a sensitive explosive (40) with a high power rating (166) and a detonation velocity of 26,500 feet (8,100 meters) per second. It can be used as an intermediary charge, but it is extensively used in detonating cords (see Chapter 4).

RDX, also called *Herogen* or *cyclonite,* was discovered by the German Henning in 1899. It is the product of the nitration of hexamethylenetetramine, and is a very powerful explosive (167) with a high velocity of detonation, 27,560 feet (8,400 meters) per second, and a moderate sensitivity of 55.

*Tetryl* is formed when dimethylaniline is nitrated. It requires careful extraction and preparation for use as an explosive, as it is powdery and toxic. It has a detonation velocity of 23,950 feet (7,000 meters) per second, a power value of 120, and a sensitivity of 70, which makes it a good intermediary charge to amplify an initiator's output and transfer it to a main charge.

HNS, *hexanitrostilbene*, is a recently developed secondary explosive that has the capability of functioning over a wide temperature range: −320°F to +525°F (−195°C to +274°C). Its detonating velocity is 22,966 feet (7,000 meters) per second with a power rating of 115 and sensitivity of 65. The high operating temperature capability of HNS has proven beneficial in weight-critical aerospace applications of pyrotechnic devices and systems that otherwise would necessitate considerable quantities of insulation if some other explosive were used.

## DEFLAGRATING PROPELLANTS

Deflagrating propellants are available in many different forms, including powders, fine and coarse; grain; spherical beads; pellets, extruded and pressed; billets, machined and cast. The different forms primarily affect their burning rate when all other parameters such as composition and density are equal. When a very high peak pressure is required in a few milliseconds, flake or granular powders are used; conversely, a low pressure for a longer period of time is obtained by using pellets or a billet of propellant. The two characteristics that all forms of deflagrating compositions have are that they all produce high volumes of gas and high temperatures.

The chemical compositions of deflagrating materials are far too numerous to elaborate here, and also, most manufacturers of pyrotechnic devices and systems have developed their own proprietary blends. A change in the formula of a proprietary propellant, to alter its deflagrating characteristics for a specific application, is another way experienced suppliers can reduce the development time and cost of a pyrotechnic device or system and thereby gain a competitive edge.

## PRIMING MATERIALS

Priming or "first fire" materials are the first pyrotechnic materials initiated in a pyrotechnic train and are used to initiate the detonation of less sensitive secondary or high explosives. Few explosives have the desired sensitivity to initiation via shock, friction, electric spark, or high temperature to make them suitable as priming explosives. They must be able to instantly release huge quantities of calories and/or shock upon initiation by an electrically heated bridgewire, the shock of a firing pin, or the flash of a laser. The main function of the priming material is to initiate a more stable and higher power generating explosive. Some of the more common priming compounds are: lead azide, lead styphnate, LMNR (lead mononitroresorcinate), KDNBF (potassium dinitrobenzofurozan), barium styphnate, and zirconium potassium perchlorate.

## TIME-DELAY COMPOSITIONS

In many pyrotechnic devices and systems it is necessary to delay a subsequent function in order to get the proper initiation sequence of two or more pyrotechnic devices. For instance, when two stages of a spacecraft booster are separated, it is not unusual to have to sever an electrical umbilical as well as the circumferential ring holding the two stages together. The ring severance system will utilize one of the many linear pyrotechnics (see Chapter 4) that are composed of secondary or high explosives, while the guillotine that severs the umbilical will be powered by a gas-generating cartridge (see Chapter 6). If both the ring severance system and the guillotine are initiated simultaneously, the ring will be severed in microseconds whereas the umbilical will require several milliseconds to be severed. This is due to the ring severance system's rapid detonating rate of 22,965 to 29,500 feet (7,000 to 9,000 meters) per second versus the 1,312 feet (400 meters) per second burning rate of the guillotine's cartridge deflagrating material. The final consequence of these combustion inequities is that the two booster stages will be structurally separated and still tethered together for a few milliseconds by the unsevered electrical umbilical. These milliseconds of hesitation are sufficient time to perturbate the booster's programmed trajectory enough to possibly cause the mission to be aborted. A simple method of sequencing the ring severance system to function a few milliseconds after the umbilical guillotine is to build into the ring severance system's initiator a pyrotechnic time-delay column between the priming mix and the detonating main output charge.

The forward end of the delay column, next to the priming mix, is composed of a heat-sensitive and fast-burning metal oxidant compound to ignite the main charge of the delay column, which is a slower-burning delay composition. The aft end of the delay column may have intermediate boosters between it and the final output charge of the device. Oxidants for the metal powder-oxidant time-delay columns are nitrate, perchlorate, chlorate, and chromate. These materials supply oxygen during combustion, without generating an excessive amount of gas, which is an important characteristic for time-delay columns since they occupy the same volume before and after burning. Otherwise, excessive gas pressure could cause premature ignition of the time-delay column. The metal constituent of the delay columns is often nickel-zirconium. The burning rate can be varied by altering the metal constituent proportions; e.g., a composition consisting of 70 percent nickel, 30 percent zirconium, and an oxidant burns at a rate of 12 seconds per inch (4.7 seconds per centimeter), whereas in a similar composition in which the percentages of the nickel and zirconium are reversed, the burning rate is 25 seconds per inch (9.8 seconds per centimeter). Other parameters that can affect the burning rate of like time-delay column compositions are particle sizes, packing densities, and ambient temperatures during storage and operation.

## LIQUID EXPLOSIVES

Nitroglycerin, mentioned earlier in this chapter, is very shock-sensitive and therefore does not enjoy widespread use. There are, however, other liquid explosives that are transported and stored as two or more liquid compounds. By themselves, these compounds are quite stable and therefore safe to handle. At the using site, they are mixed and used almost immediately. Liquid explosives in commercial use today are the primary result of privately funded research and development programs and therefore these formulas are proprietary. Chapter 5 does discuss a unique underwater application of liquid explosives in conical shaped charges.

chapter 3

# Initiators, Detonators, and Primers

Any device that receives and processes some unique energy signal, i.e., electrical (low and high voltage), explosive, ballistic hot-gas, mechanical, or laser, to cause an exothermic reaction in one or more chemical elements or compounds in the device (or in contact with it) is an initiator. The output is heat, shock, pressure, or a combination of these. Each type of initiator has implicit commitments as to energy source and transmission lines for that particular energy. Figure 3-1 illustrates the numerous energy sources mentioned above, and their transfer media to the initiator. The impacts of the energy source and transfer media have to be considered in each initiator application. At the initiator input and output interfaces, special devices may be needed to interconnect with adjacent components. For example, the input interface at the initiator will need connectors for electrical cable, pyrotechnic booster tips and couplings for detonating cords (see Chapter 4), couplings for hose, etc. The output interface will be concerned with debris control devices, gap control, etc. An initiator processes an energy input, transducing it to an output with the characteristics to ignite, prime, or detonate.

## ELECTRIC INITIATORS

In the beginning all electric initiators looked like the typical blasting cap, as shown in Figure 3-2. They resemble a pencil-diameter metal can with two insulated wire leads beaded in the open end. Inside, a resistance wire across the lead ends is heated by an electric current, raising the temperature of a heat-sensitive priming mix (see Chapter 2) beyond its autoignition temperature. The firing of this small amount of priming material starts a train of progressively less sensitive charges leading to the main output charge. Even today, blasting caps, being less expensive than any other type of electric initiator requiring an electrical connector, are used in the initial phases of pyrotechnic device development testing.

Electric initiators are distinguished from an amorphous group called

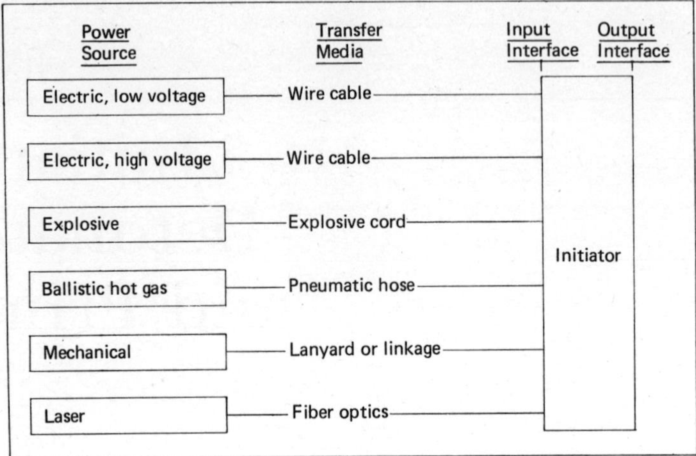


**FIG. 3-1.** Initiaton power sources and transfer media.

"nonelectric." The distinction is helpful in safety considerations where the electric initiators have the susceptibility to accidental firing by their absorption of radio frequency (RF) energy. In an RF field the leads and wiring can act as efficient dipole-receiving antennas. The nonelectric initiators are immune to RF energy.

The effort to make electric initiators safe in an RF-saturated environment, in which some are used, has led to the 1-ampere, 1-watt initiator classification. Units with this capability can dissipate either 1 ampere of current or 1 watt of power delivered through their bridgewires for five minutes without firing or being degraded for future use.

A second hazard seen by electric initiators is static electricity. An operator can build an electric charge on his or her person. The kind of clothing worn, the type of floor covering, and atmospheric conditions affect the static buildup. Static electrical charges can easily be discharged by simply touching an object of lower electrical potential, and if that object happens to be an

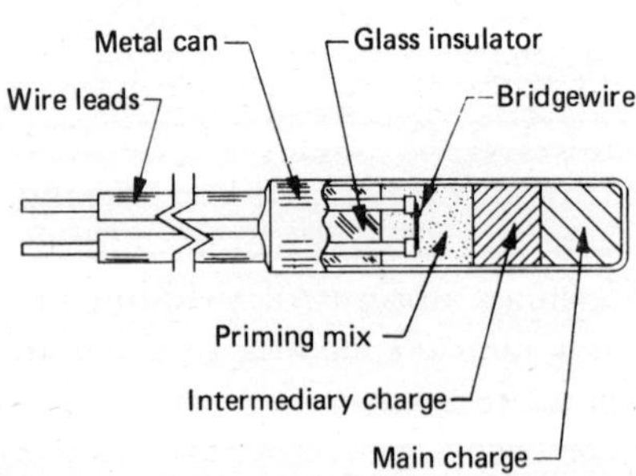


**FIG. 3-2.** Electric, low-voltage, hot-wire blasting cap.

electric initiator, an accidental initiation can occur. The general requirement that electric initiators withstand 25,000 volts discharged from a 500-picofarad capacitor without firing derives from the charge that could be built up on the human body.

Static charges can also build up on plastic sheets, particularly when they are in the process of being moved. For example, the act of uncovering a shrouded spacecraft is potentially hazardous, if electric initiators are onboard, because static electricity can be generated during the process of moving the plastic shroud. Flowing matter, either granular solids or liquids, will build up a static charge. Low humidity and particulate content in the air can enhance the generation of static electricity.

Grounding of operators and common-ground bonding of all conductive components, as well as humidity control and proper selection of clothing, are proven protective measures. Plain cotton cloth is best in defeating static electricity. Synthetic fabrics are generally to be avoided in handling electric initiators as well as pyrotechnics in general.

## Electric Hot-Wire Initiator

The hot-wire initiator is a high-fidelity version of the blasting cap that is also manufactured under a quality assurance and control program (see Chapter 8) to increase its reliability. As shown in Figure 3-3, it uses a bridgewire heated by an electric current until a heat-sensitive priming mix reaches its autoignition temperature. The priming mix is frequently painted or "buttered" directly onto the bridgewire. An intermediary charge (or several) may be used in the progression to the output charge. The electrical contact pins are sealed into a glass insulator at the input end. A closure is soldered, welded, or epoxy-bonded to the initiator housing output end. The insulator and closure provide hermetic seals to the housing, which is also configured

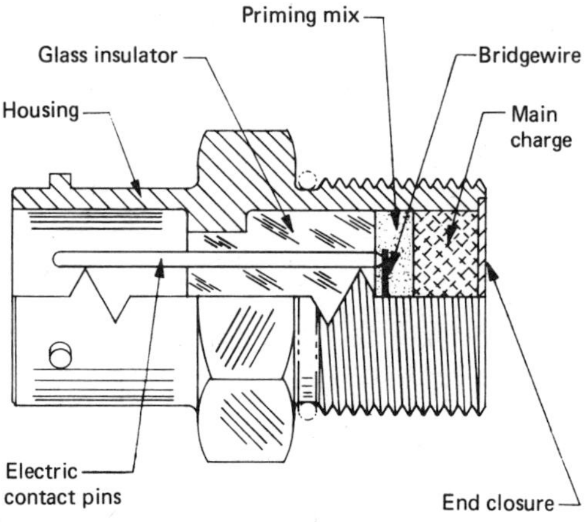


**FIG. 3-3.** Electric, low-voltage, hot-wire initiator.

for electric connector threads or bayonets, flanges, threads, torquing flats, etc., for the conveniences of installation.

For many years, the heat-sensitive priming material was black powder. Mercury fulminate saw extensive use but it fell into disfavor because its shelf-life deterioration makes it unusable in the tropics (3 years at 35°F or 1.7°C will render it unable to detonate). The United States stopped using it about 1930 and switched to lead azide. Mercury fulminate, even in good condition, is so sensitive to heat and shock that only small quantities can be handled safely. It is no longer used as a priming mix. Most initiator manufacturers have developed their own proprietary mixes to meet the 1-ampere, 1-watt challenge which began in the 1950s.

## Electric Exploding Bridgewire Initiator (EBW)

The exploding bridgewire (EBW) is the second type of electric initiator. It differs from the hot-wire type primarily in the bridgewire's purpose. The exploding bridgewire does just that—it explodes under a massive surge of current dumped into the circuit in a few microseconds by a discharging capacitor. The bridgewire explosion is similar to the filament of an incandescent light bulb flashing when the light switch is turned on and the bulb flashes and burns out. In the process of burning out the filament, the quantity of heat generated is several times that of the normally glowing filament for a like period of time. The shock and heat of the EBW is sufficient to initiate a secondary or main pyrotechnic charge (see Chapter 2) directly, without the interposition of a sensitive primary pyrotechnic mix. No primary charge material is used in the EBW initiator.

The illustration in Figure 3-4 shows only one electric contact pin, while actual units may have either one or two, with a spark gap in only one of the legs. The purpose of the spark gap is to interrupt all but relatively high

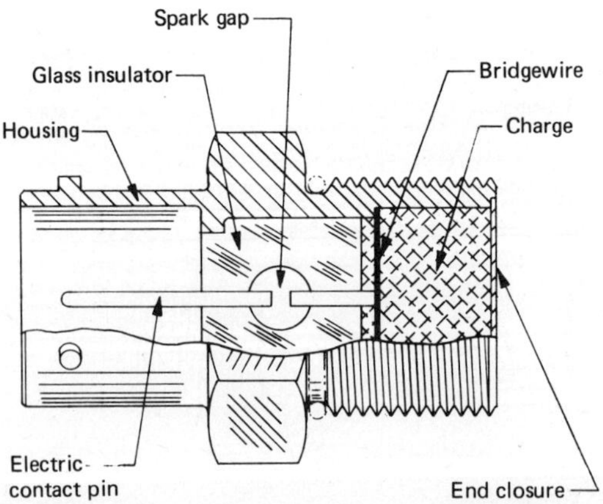


**FIG. 3-4.** Electric, high-voltage, exploding bridgewire (EBW) initiator.

voltages, i.e., everything in excess of 600 volts. The gap protects against accidental exposure to the prevalent 230 volts AC, particularly on shipboard installations. The gap is selected to provide a safety margin above the peak values of the prevalent current in the initiator's operation environment.

The best pyrotechnic charge for EBW initiation is a fine powder surrounding the exploding bridgewire. The finer the powder, the better. A hard, compacted mass is not effective. The fine, loose particles crushing against each other under shock provide the effective agent of detonation. A hard mass tends to fragment rather than initiate.

EBW initiators have not gained wide use because of test requirements to prove their resistance to RF energy. Requirements stipulate that the 1-ampere or 1-watt power levels will be introduced to the bridgewire directly, i.e., bypassing the spark gap. EBW initiators will pass the 1-ampere current test but generally fail the 1-watt exposure test. The failure mode is that the bridgewire heats up, causing the pyrotechnic charge next to it to deteriorate. Initiators sometimes dud when they are subsequently called on to fire. Efforts to separate the charge from the bridgewire, by either an air gap or an insulator, decrease the initiator's reliability compared to intimately placed secondary charges.

EBW initiators have several special requirements which restrict their use. They require a heavy power load that must be provided from a special firing unit. Charging its capacitor takes some finite time depending on the electrical characteristics of the firing unit. The firing circuitry for an EBW initiator generates powerful RF interference, since it bursts into action with a spike of energy with unavoidable secondary emissions. The special low-impedance electric cabling from the firing unit to the EBW initiator is a radiating source of possible interference with other on-board electronics. For this reason the firing cable is generally shielded with metal braid to surpress RF interference, thereby increasing system cost and weight.

Because electrical voltage line losses are significant over long distances (more than 9 feet or 3 meters), even with special low-impedance cable, it is usual practice to keep the firing unit close to the EBW initiator. Installation requirements may therefore require special triggering control cable from a remote firing location, e.g., an aircraft cockpit or spacecraft crew module to the firing unit. EBW initiators are not the first choice for use where weight and power requirements are limiting factors.

## EXPLOSIVE-ENERGIZED INITIATORS

Explosive-energized initiators come in many configurations; however, only the three most popular will be presented here. They may be located at the boostered terminus of the main line of an explosive train or one of its branches. A special requirement of an explosive interface occasionally encountered is to maintain a pressure seal within an explosive train after passage of the pyrotechnic stimulus. This is accomplished by a device called a "thru-bulkhead initiator." Finally, the high-energy, radial output of a detonating cord can be tapped many times without having to splice into the cord itself to activate an initiator.

## SMDC Initiator

The axial output of a shielded mild detonating cord (SMDC) (see Chapter 4) booster tip is used to start a new function as shown in Figure 3-5. If a time-delay/one-way transfer function is desired, the SMDC axial detonation output can be processed within the explosive-energized initiator to begin the new function. A dimpled, cup-shaped, firing pin is mounted behind a percussion primer. The detonating output from the SMDC causes the cup to "undimple," thereby forcing the protruding firing pin to strike the primer. The backside of the firing pin cup is sometimes filled with a special rubberized material to soften the output shock of the SMDC. This type of explosive-energized initiator is usually an integral part of the next functioning device where the output of the percussion primer initiates a priming mix and subsequently progresses on to the final output function of the device, i.e., gas pressure, flame, shock, etc. It is to be noted that this type of initiator requires the termination of the initiating-explosive train at this device, otherwise the input SMDC to this explosive-energized initiator must be from a branch of the main SMDC train.

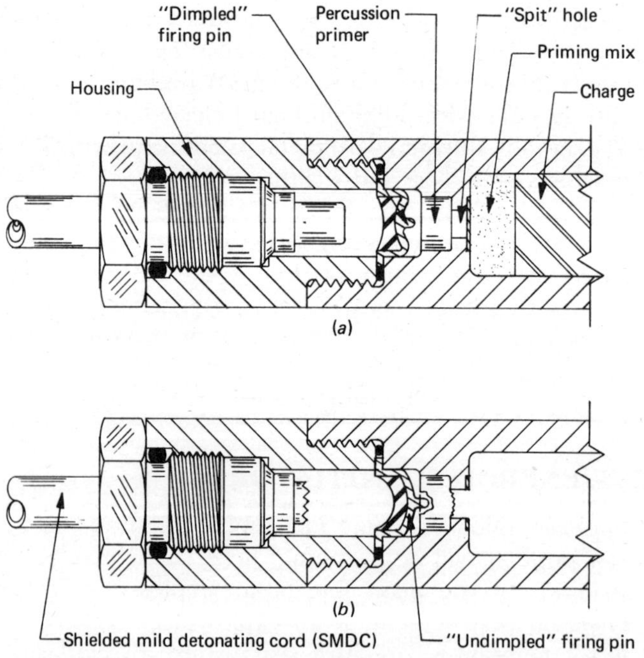


**FIG. 3-5.** Explosive-energized (SMDC) initiator. (*a*) Preinitiation. (*b*) Postinitiation.

### Thru-Bulkhead Initiator (TBI)

A requirement often emerges to maintain a pressure seal between the input stimulus and the output of the initiator's main pyrotechnic charge. The output charge may be a gas-generating propellant that drives a piston in a thruster or some other type of stroking action. Figure 3-6 illustrates a thru-bulkhead initiator (TBI) that could be installed in a thruster at the threaded end. The input stimulus enters the initiator from the SMDC tip. The explosive output from the tip initiates a donor pyrotechnic high-explosive charge pressed into the housing. The exploding donor charge generates a shock wave through the initiator's housing toward an acceptor charge similarly pressed into the housing a short distance away, approximately 0.05 to 0.10 inch (1.27 to 2.54 millimeters). The acceptor charge is thus initiated by the shock wave from the donor charge. The acceptor charge in turn initiates a priming mix and finally the main output charge of the TBI. The characteristics of the housing material have to be generally those of steel, otherwise the ductility of a softer material, such as aluminum, tends to attenuate the shock waves traveling through the bulkhead. The bulkhead could be made thinner, but then it may not be strong enough to contain the high pressures generated by the initiator's main output charge.

### Detonating Cord Initiator

Figure 3-7 illustrates an explosive-energized initiator that enables an unconfined detonating cord (see Chapter 4) to be uninterrupted (unbroken) and yet connected to an unlimited number of initiators. The explosive weight of detonating cords used to set off a train of initiators may be 50 grains per foot (152.4 grains per meter) or larger. The cord is installed within the initiator by removing a cap and ferrule. The cord is placed at the base of an exposed slot against a percussion primer. The aforementioned ferrule is placed to straddle the cord, and the cap retains the cord against the primer. As the detonating shock wave travels along the cord and through the

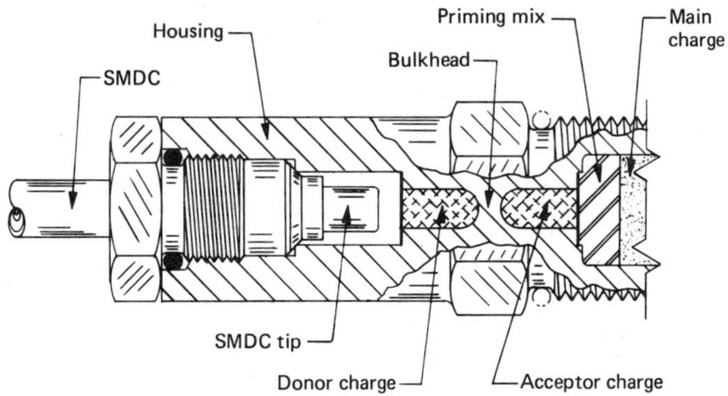


**FIG. 3-6.** Explosive-energized thru-bulkhead initiator (TBI).

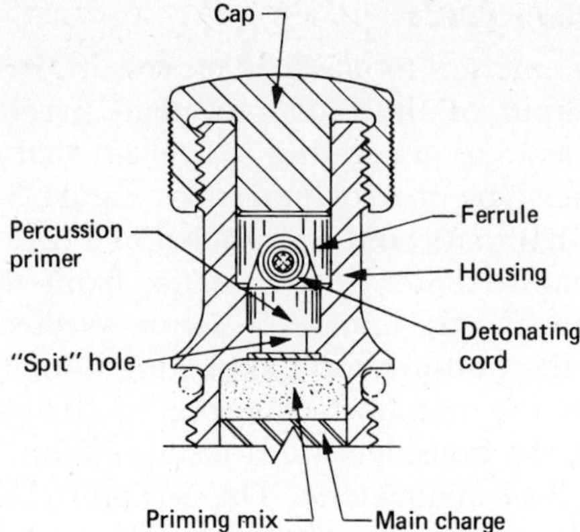


**FIG. 3-7.** Explosive-energized (detonating cord) initiator.

initiator, the shock wave fires the percussion primer that in turn initiates a priming mix next to the initiator's main output charge.

## BALLISTIC HOT-GAS-ENERGIZED INITIATOR

The ballistic hot-gas-energized initiator accepts a high-pressure surge from a pyrotechnic gas-generating device that can be delivered to the initiator by either a rigid metal tube or flexible hose. The latter is most apt for interconnections of two stations that may have to move with respect to each other (e.g., an adjustable crew-escape ejection seat). Figure 3-8 depicts the major components of such an initiator. The high-pressure hot gas creates a force behind a piston that has a firing pin on its opposite end. The force

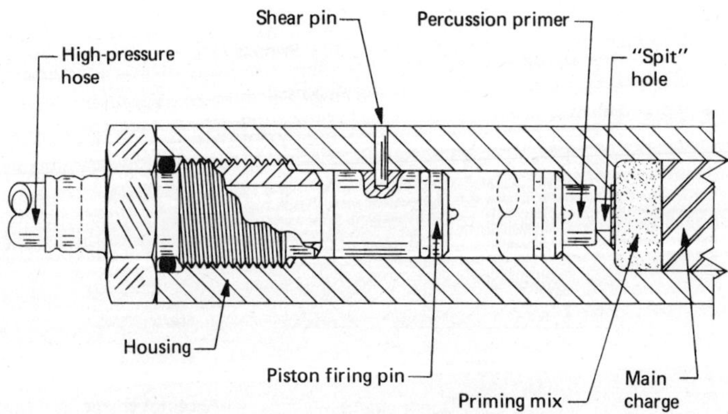


**FIG. 3-8.** Ballistic hot-gas-energized initiator.

shears a piston-retaining pin, allowing the piston to move a short distance before striking a percussion primer. The output of the primer initiates a priming mix nearby and subsequently progresses on to the output function of the device, i.e., gas pressure, flame, shock, etc.

## MECHANICAL-ENERGIZED INITIATORS

The most frequently used mechanical initiators are those that require the user to either directly or remotely pull a handle that simultaneously retracts a firing pin and compresses a spring behind it. At some point in the travel the retracting force is suddenly removed and the firing pin fires the initiator. A unique application of hydrostatic pressure is utilized to arm and fire an underwater initiator in a somewhat different type of mechanical-energized initiator.

### Manually Energized Initiator

Spring-loaded firing pin types of initiators have widespread use in aircraft crew-escape systems. The external handle of Figure 3-9 is often the D ring located between the pilot's knees on ejection seats. The safety pin is usually removed at the beginning of the flight and reinstalled at the conclusion. In an emergency, the pilot starts the ejection sequence by pulling the D ring. As the D ring is pulled, it retracts the firing pin and compresses a spring behind it. When the sear interface between the D ring and the firing pin is moved to the outside of the initiator housing, the firing pin is released from the D ring, and the compressed spring rams the firing pin against a percussion primer. The primer in turn initiates a nearby pyrotechnic priming mix and

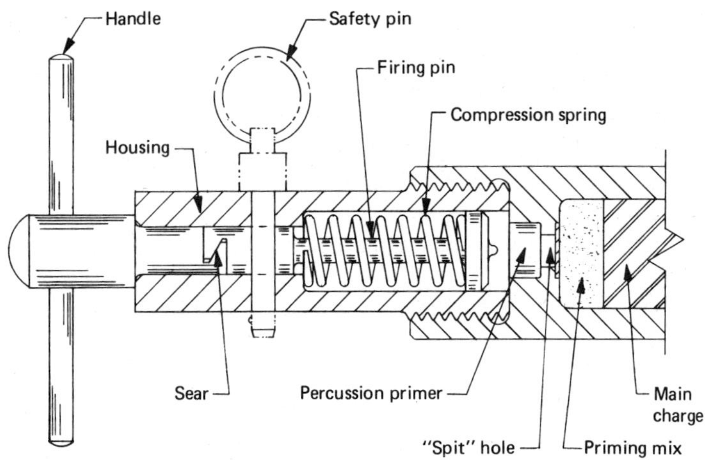


**FIG. 3-9.** Manually energized initiator.

subsequently progresses on to the output function of the device, i.e., gas pressure, shock, flame, etc.

**Hydrostatic Initiator**

An initiator designed to utilize the increasing pressure associated with increasing water depth is a hydrostatic initiator. This generally requires two sequenced functions, i.e., arm and fire, to effect an initiation. Sound fixing and ranging (SOFAR) bombs employ such an initiator. SOFAR bombs are bomblike explosive devices that are dropped by ships and airplanes to accurately fix a location within a large body of water such as a ship sinking, oil slicks, reefs, and sandbars. Underwater direction-sensitive microphones (hydrophones), permanently located in all the oceans, seas, and large lakes of the world, are being constantly monitered for the explosive sound of SOFAR bombs. The sound of an underwater explosion travels great distances, and the greater the depth of the initiated SOFAR bomb the greater the distance the sound will travel. The sensitive receiving equipment can accurately determine the direction from which the explosive sound came; if the direction of a particular sound is plotted from several receiving hydrophones, that point at which two or more direction lines cross is the exact location of the SOFAR bomb.

Figure 3-10*a* shows a typical hydrostatic initiator with the two dimple springs located in the arming and firing circuits in Figure 3-10*b*. The dimple springs are thin aluminum disks with raised dimples in their centers. As the SOFAR bomb sinks, water pressure over the dimples exerts pressure that tries to turn the dimples inside out. This inverting motion is not gradual, but more of a snap action at the terminal depth. The pressure required to "trigger the dimple" is a function of the dimple spring material properties, thickness, and dimple geometry. Usually the two dimple springs of a hydrostatic initiator are identical, with only their thicknesses being slightly different. The greater the thickness, the higher the pressure required to trigger it. Therefore the arming dimple spring may be selected to trigger at an ocean depth of 1,500 feet (457 meters), where the hydrostatic pressure is 663 pounds per square inch (46.6 kilograms per square centimeter). Directly behind the arming dimple spring is a slider containing a pyrotechnic primer. When the arming dimple spring triggers, it moves the slider, Figure 3-10*c*, just enough to align the primer with a firing pin in the firing circuit.

The firing dimple spring may have a thickness designed to trigger at a depth of 2,500 feet (762 meters), where the hydrostatic pressure is 1,105 pounds per square inch (77.7 kilograms per square centimeter). The snap action of the firing dimple spring rapidly drives the firing pin, Figure 3-10*d*, into the primer that in turn initiates a pyrotechnic booster, containing a priming mix and finally an intermediary charge whose output is capable of detonating the main charge of the SOFAR bomb.

These parameters of the illustrated hydrostatic initiator are only hypothetical, and the arming and firing dimple springs may be designed to trigger

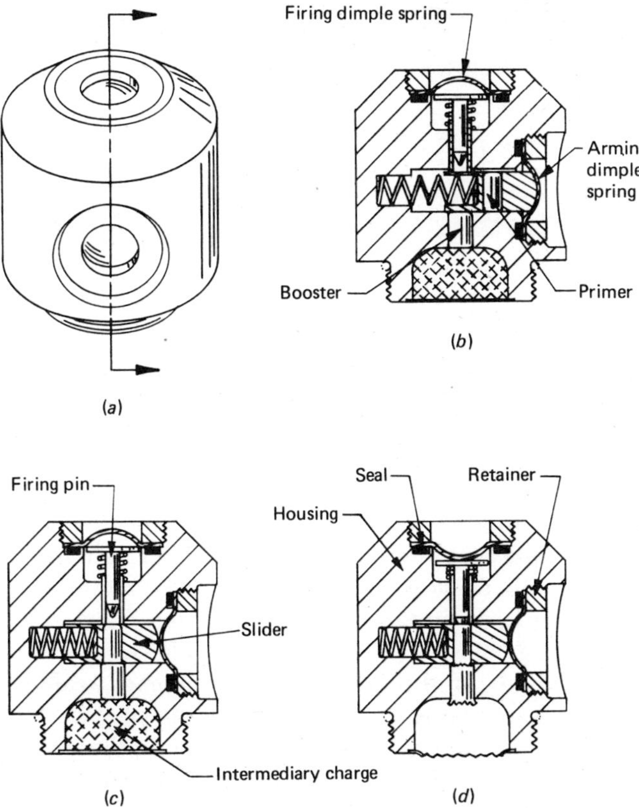


**FIG. 3-10.** Hydrostatic initiator. (*a*) Initiator. (*b*) Initial configuration. (*c*) Armed configuration. (*d*) Fired configuration.

at any depth beyond 100 feet (30 meters). Differential arming and firing depths are usually not less than this because the dimple spring trigger parameters become difficult to control, and the firing dimple spring may fire at a shallower depth than the arming spring.

## LASER-ENERGIZED INITIATOR

The development of the laser principle of coherent light generation has resulted in a entirely new concept in pyrotechnic initiation. A laser generates a highly collimated beam of nearly coherent light capable of initiating pyrotechnic compositions. The use of laser light at the explosive interface eliminates the conventional bridgewire and all electrical wiring to the initiator. Because of this unique configuration, laser initiators are immune to accidental firing by electromagnetic fields, electrostatic discharges, or stray electrical energy. Figure 3-11 shows a simple laser-energized initiator consisting of an

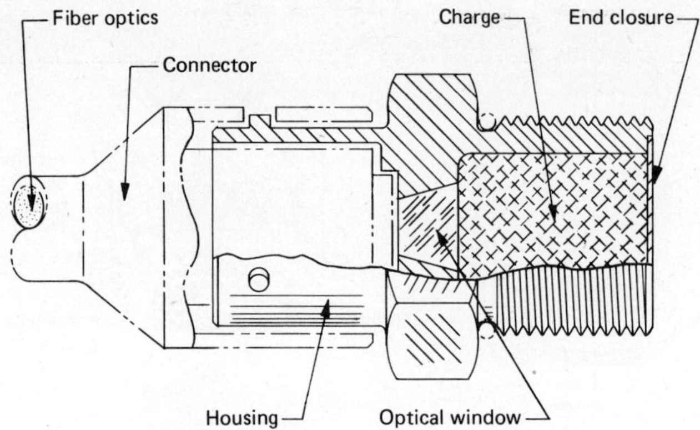


**FIG. 3-11.** Laser-energized initiator.

initiator housing, an optical window, and a pyrotechnic charge retained by a closure disk. Fiber optics are used to transmit the laser light into the initiator. Additional auxiliary fibers allow the optical path to be monitored at any time before firing. The optical window can be constructed from almost any high-quality optical glass, as most glasses have acceptable transmission characteristics at the wavelength of the laser. The problem of sealing the optical window into the initiator body can be solved by the proper selection of the body material that is compatible with glass. An optical window can be designed to withstand pressures up to 50,000 pounds per square inch (3,500 kilograms per square centimeter), which is well within the operating range of most other types of initiators. Due to the high development costs of the other laser components, the laser-energized initiator has found only limited laboratory use. Perhaps as manufacturing costs abate, more interest will be shown in this unique initiation concept.

## DETONATORS

A detonator delivers shock output to induce high-order detonation, i.e., wave velocities greater than 16,400 feet (5,000 meters) per second into an adjacent acceptor charge. Any of the initiators described previously in this chapter can be made into a detonator by simply making the main output charge a high or secondary explosive. Some require an intermediary explosive just ahead of the main charge. Figure 3-12 illustrates two types of cartridge detonators that have evolved. The conventional configuration is capable of transferring a detonating shock wave when the acceptor is oriented radially or axially to the output tip of the detonator. This is due to detonating shock waves radiating from the surface of the explosive.

End-firing cartridge detonators differ primarily in having a thicker cylindrical wall on the output tip to increase the axial output-shock intensity.

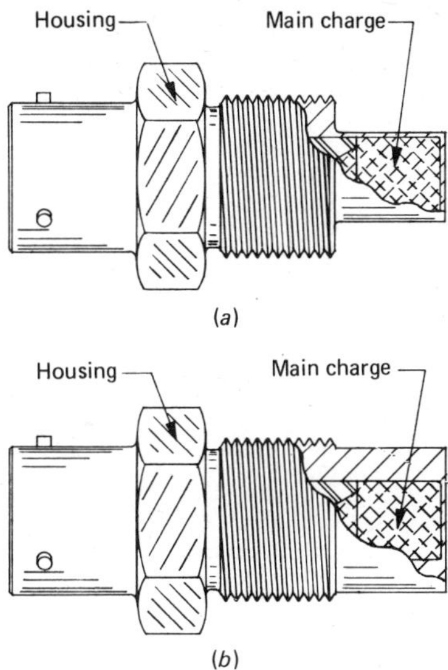


**FIG. 3-12.** Detonator outputs. (*a*) Conventional detonator. (*b*) End-firing detonator.

Hemispheres could replace the flat, circular ends on the output tips. This configuration, however, would also provide a much larger surface from which the same amount of shock-wave energy could radiate, resulting in a much lower shock-wave concentration and lower reliability to detonate the acceptor charge.

## PRIMERS

Primer is the name given to a group of mechanically or detonation-shock-fired initiators. The mechanical firing device is a firing pin. The two most commonly used primers are *percussion* and *stab* primers. Figure 3-13 illustrates the main differences between the two types.

### Percussion Primer

Percussion primers are flame-producing initiating devices that are activated by a combination of friction and impact (mild shock). They are mainly used to ignite the smokeless powder in gun cartridges, black powder, time-delay trains, and the intermediary explosive in detonators. They consist of a drawn, thin-walled, tubular housing containing an impact-sensitive explosive charge, a closure disk, and an anvil. When a blunt-end firing pin dents the housing, it squeezes the charge between the dent and the anvil, causing it to ignite. Machined into the anvil are axial slots that permit the ignited propellant to bypass the anvil and rupture the rear end of the primer case. The installation of a percussion primer has some specific requirements that include rigid

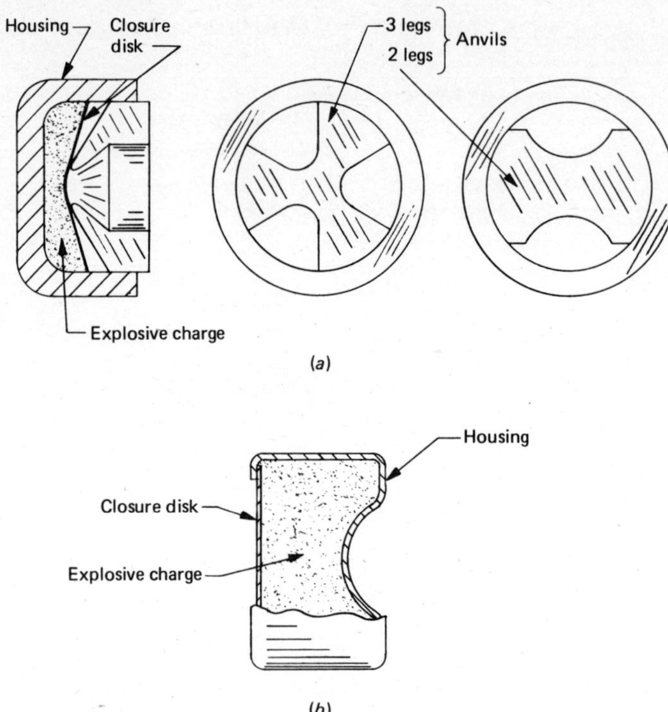


**FIG. 3-13.** Primers. (*a*) Percussion primers. (*b*) Stab primer.

seating, e.g., bonding, pressing, or staking in place, and a short open distance, a "spit" hole, between the output side of the primer and the thin closure disk at the entrance to the pyrotechnic chamber. The "spit" hole (see Figures 3-5, 3-7, 3-8, and 3-9) provides unobstructed clearance for the burning propellant and metal casing fragments to accelerate to optimum momentum (this phenomenon is called "brisance"), resulting in higher reliability to initiate the adjacent priming mix. The capability of the percussion primer to contain all the gases and high pressures on the firing-pin side is one of the main advantages of this type of primer.

## Stab Primer

Stab primers also use drawn, thin-walled, tubular housing, a compressed primer charge, and a closure disk, but no anvil. The firing pin of a stab primer is much more pointed than that of the percussion primer. It pierces a hole in the thin closure disk and the explosive charge, causing it to ignite.

From the stab primer has evolved the stab detonator by the addition of intermediate and main charges after the priming mix. Some stab detonators have even included a time-delay column. Stab detonators have found their greatest application in munitions warhead initiation and will not be discussed any further here.

Typical stab primer and stab detonator initiation parameters are: an all-fire sensitivity of 1.25 inch-ounces (900 millimeter-grams) kinetic energy; i.e., a 0.25-ounce (7.09-gram) firing pin with a 26-degree taper angle and a 0.007-inch (0.178-millimeter) flat face dropped from a height of 5 inches (127 millimeters) above the initiating face of the primer.

chapter 4

# Linear Pyrotechnics and Severance Systems

The pyrotechnic materials addressed in this chapter will be restricted to high or secondary explosives and are not to be confused with the familiar sparking and smoking type of safety fuse used to initiate dynamite. Whereas the safety fuse will burn at a rate of a few feet or meters per minute, the high explosives detonate at a rate in excess of 16,400 feet (5,000 meters) per second.

Although the final geometries of the linear explosives may vary, they all generally start out as a metal tube (or sheath), e.g., lead, aluminum, silver, or copper of some manageable length, approximately 10 feet (3 meters), and an outside diameter between 0.38 and 0.75 inches (1 and 2 centimeters), whose inside diameter is approximately 50 percent of the outside diameter, depending, in part, on the characteristics of the final product. One end of the tube is capped and the tube is incrementally packed, under high pressure, with a granular high explosive such as PETN, RDX, or HNS (see Chapter 2). After the tube has been completely filled and capped, it is subjected to a drawing process that may include reduction dies and many proprietary processes and procedures that are, understandably, selfishly guarded. The units used to define the quantity of explosive per given length (core loading) are the grains per foot (grams per meter) method. The dominant core loadings are between 2 and 500 grains per foot (0.43 and 106 grams per meter), although for special applications they have been drawn as low as 0.2 grain per foot (0.04 gram per meter) and as large as 2,000 grains per foot (425 grams per meter).

The drawing process increases the length of the tube and the explosive within, but it can either retain the original circular shape or it can alter it to almost any desired configuration, as shown in Figure 4-1. If the final configuration remains circular, it could have application as one of several types of stimuli transfer devices or it can be used as an efficient structural severance device, all of which will be discussed later in this chapter. Linear explosives with round and rectangular cross sections have found broad

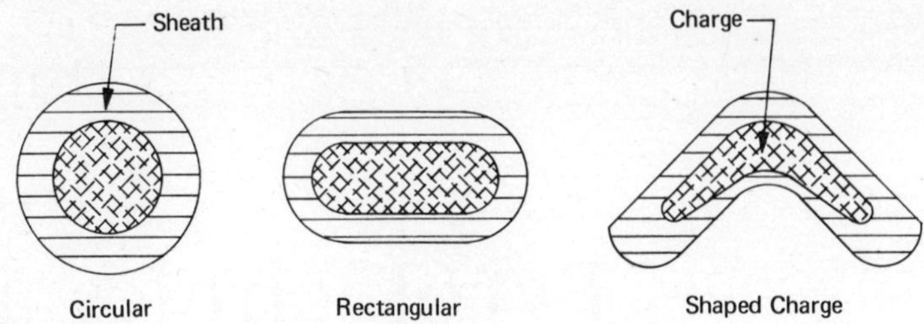


**FIG. 4-1.** Typical linear explosives.

application in the aerospace industry in severing large umbilicals that may contain wires, tubes, cables, hoses, or any combination of these.

By modifying the initial circular shape to a specific chevron design, an extremely efficient structural severance device, the linear-shaped charge, is produced. Chapter 5 delves into the principles of the shaped charge and presents many variations and applications of it.

An interesting comparison of the energy-dispersion patterns of three different cross sections of linear explosives with identical core loadings, core materials, and sheathings is shown in Figure 4-2. Three 7.5 grains per foot (1.6 grams per meter) circular, rectangular, and chevron (shaped charge) specimens were sandwiched between two 8-inch (20.3-centimeter)-thick

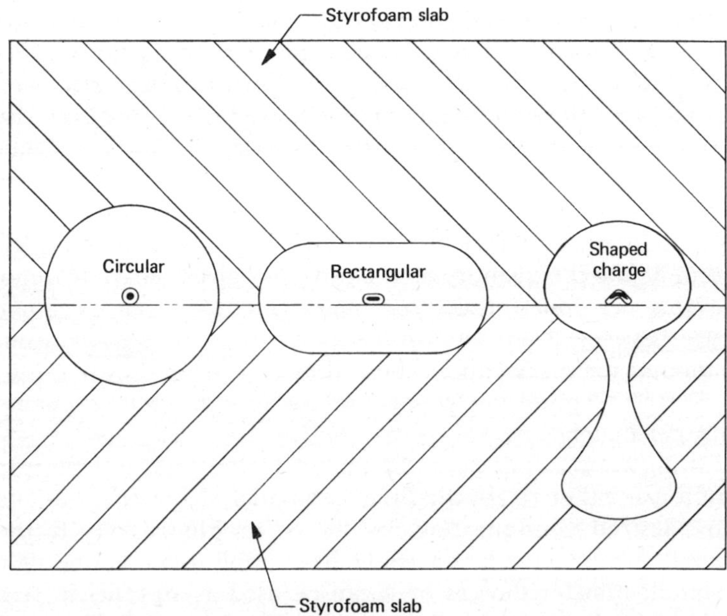


**FIG. 4-2.** Energy dispersion patterns of linear explosives.

blocks of medium-density styrofoam and fired. The areas of all three cavities were within 5 percent of each other, which is well within the core loading manufacturing tolerances.

## MILD DETONATING CORD (MDC)

The origin of the term "mild detonating cord" (MDC) is obscure and slightly confusing, since it is not clear what the originator was comparing the detonating cord with to refer to it as "mild." Nuclear energy is one of the few energy sources more powerful than high or secondary explosives.

An unusual application of MDC was to propel two pairs of opposing guillotine blades, in one assembly, through two tandem umbilicals. Such was the design of the Apollo command to service module (CSM) umbilical guillotine shown in Figure 4-3*a*. The umbilical was composed of 2,300 wires of various cross sections and eight steel tubes, between 0.25 and 0.63 inch (6.4 and 16 millimeters) outside diameter. The wires and tubes were about equally divided and potted into two 0.88- by 9-inch (2.2- by 22.9-centimeter) umbilicals. As part of the CSM's separation sequence, just prior to the command module's fiery reentry into the earth's atmosphere, the two umbilicals were severed by propelling opposing pairs of guillotine blades toward each other and through each umbilical routed through openings provided in the guillotine's housing.

Each of the four blades (see Figure 4-3*b*) was propelled by dual strands of 20 grains per foot (4.25 grams per meter), lead-sheathed MDC with RDX explosive core and two RDX end boosters of 19 grains (377 milligrams) each. The boostered cord assembly was potted into a matching groove on top of each blade. The potting material was removed a short distance in the middle of the blade to expose the dual MDCs to the output of the initiating detonator. A unique feature of the blade design was that the cutting edge

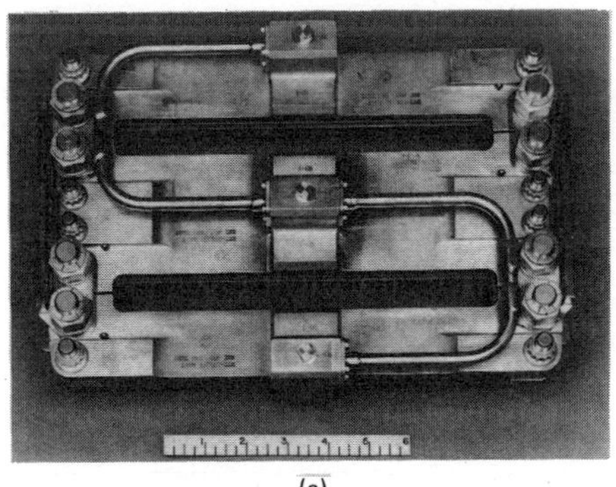
(*a*)

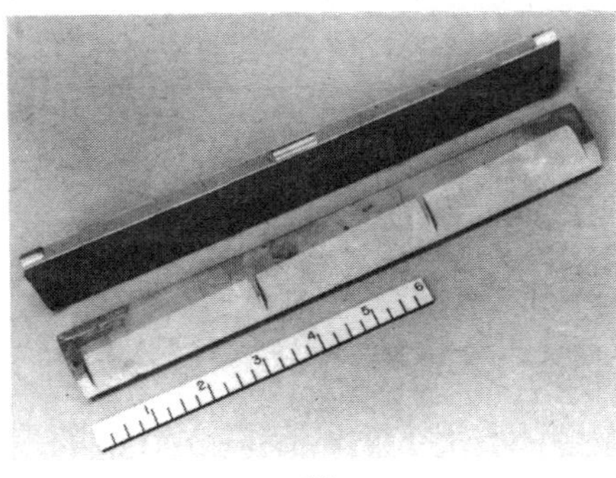
(*b*)

**FIG. 4-3.** Apollo command-to-service module (CSM) umbilical guillotine. (*a*) Assembled guillotine. (*b*) Two of four guillotine blades.

was not centered along its longitudinal axis. Instead, the edge was laterally displaced 0.04 inch (1 millimeter) at each end of the blade on opposite sides of the centerline. This feature assured that the opposing blades would intersect and decelerate each other, after severing the umbilical, instead of camming alongside each other and possibly flying out of the guillotine housing.

Spacecraft and high-flying aircraft must maintain an earthlike interior atmosphere in a near-vacuum outside environment. Cabin pressure leakage through emergency crew-escape door seals is a constant concern. An escape door must be installed from the outside of the cabin with mechanical latches pushing the door against the seals. In the event of an emergency escape, these same latches must all function simultaneously to prevent jamming and allow the door to be jettisoned cleanly. Modern aircraft and flight-test manned spacecraft have alleviated this leakage concern by eliminating the door and associated latches by designing the structure, where the door would have been, to accommodate a linear pyrotechnic severance system. The following example will illustrate two completely different concepts of linear pyrotechnic severance systems utilized in the same emergency hatch jettison system.

On the space shuttle orbiter, the crew module is suspended within the outer fuselage by numerous turnbuckles. This double-hull concept minimizes

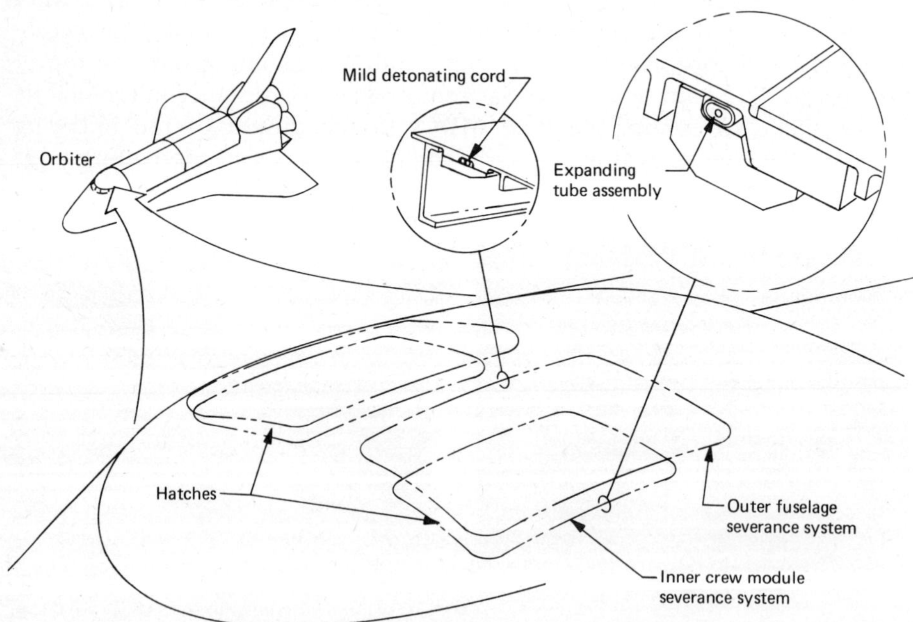


**FIG. 4-4.** Space shuttle orbiter's jettisonable hatches.

thermal shorts from the hot outer fuselage surface and the crew module during reentry from space. During flight testing, the two-man crew have the capability of escaping by first severing and jettisoning two hatches from the orbiter's double hull and then jettisoning themselves via ejection seats. Figure 4-4 shows the relationship of the two jettisonable hatches and the two panels within each hatch. Unconfined (meaning no attempt is made to confine the flame, smoke, or debris) MDC consisting of dual strands of silver sheather 35 grains per foot (7.4 grams per meter) of HNS explosive is used to sever a hole in the outer fuselage structure. The severance system that cuts a hole in the crew module is confined and will be described later in this chapter.

The installation of the dual MDCs into the outer fuselage is accomplished by machining a groove into the inside surface of the aluminum structure deep enough to install the dual MDCs as shown in Figure 4-5. Voids between the redundant strands of MDC are filled with a potting compound. The filled groove is then covered with a strong contoured aluminum charge holder (frame) attached by closely spaced bolts. The charge holder must be substantially stronger than the opposing structure being severed, otherwise it would be blasted loose from the panel when the dual MDCs are detonated. Development tests were conducted, at the minimum expected operating

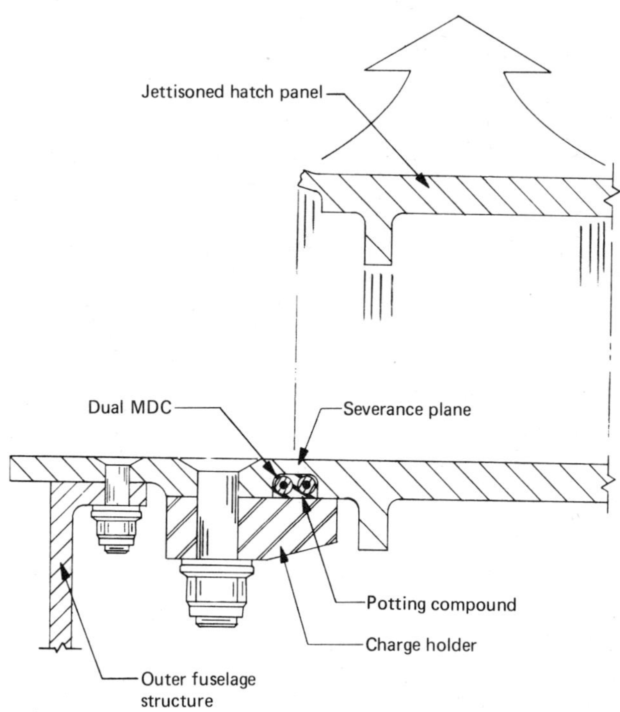


**FIG. 4-5.** Orbiter hatch, outer panel, unconfined severance system.

temperature, to determine the minimum explosive core loading of a single MDC strand to sever a test panel manufactured to maximum thickness tolerances in the severance area. These tests determined the core loading to be 32.5 grains per foot (6.9 grams per meter). This minimum core loading is then increased 7.5 percent to compensate for the minimum tolerance of the ±7.5 percent variation allowed from nominal core loading during manufacturing. This performance margin test demonstrates the selected MDC's capability to sever a panel even when several of the most influential parameters are deliberately biased to their most detrimental tolerance extremes. Conversely, the safety test is a demonstration that when the parameters are biased in the opposite extremes, the orbiter will not be blown to pieces. Here the test is conducted at the maximum expected operating temperature, using dual MDCs, each with a 7.5 percent explosive core overload to 37.5 grains per foot (8.0 grams per meter), potted into a test panel machined to minimum allowable severance section thickness. Here the pass/fail criteria include the ability not only to sever the panel but also not to create excessive debris that could be hazardous either to the crew or to the successful jettisoning of the hatch followed by the crew's ejection through the severed hole in the panel. Figure 4-6 is a photo sequence of one of the orbiter's outer panels being severed by the dual unconfined MDCs. The

(a)

(b)

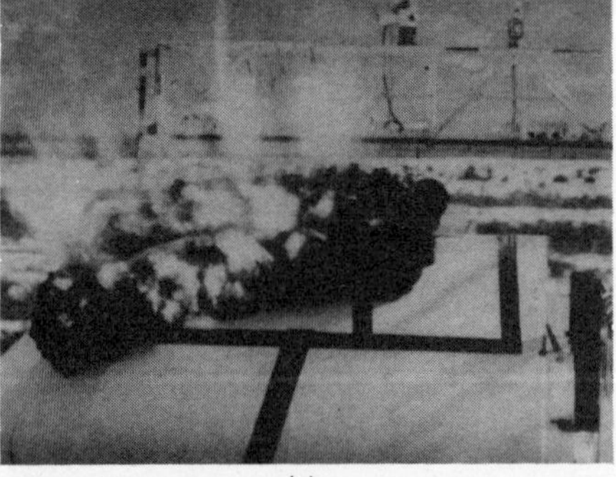

(c)

**FIG. 4-6.** Orbiter hatch, outer panel severance system test. (*a*) Initiation begins at aft lateral edge. (*b*) Panel severance is nearly complete. (*c*) Panel is completely severed.

initiation block is located in the middle of the aft lateral edge of the panel. The ends of the MDCs terminate in HNS boosters that are brought together to form a loop, and each booster is initiated with a detonator. The photos dramatically illustrate the tremendous energy of high explosives which will enable the reader to understand why the inner panel, on the top of the crew module, is severed by a confined linear pyrotechnic severance system.

Inside the crew module the astronauts' heads are as close as 14 inches (35.5 centimeters) from the edges of the panels containing the pyrotechnic severance systems. This severance system, known as the expanding tube assembly (XTA), also utilizes MDC, except in this application, shown in Figure 4-7, a single strand of MDC was used instead of two as in the outer fuselage panel severance system. The reason for this change will be explained later in this discussion. The principle of the XTA severance system is based upon *Pascal's law,* which states that pressure applied to a confined fluid is transmitted equally to all surfaces confining the fluid. The fluid in the XTA system is the flattened lead tube. When the MDC is detonated, extremely

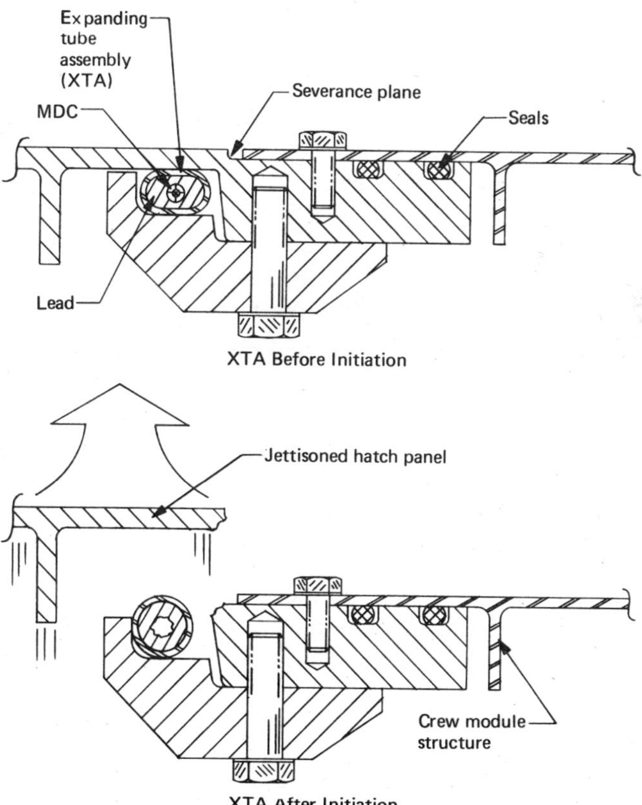


**FIG. 4-7.** Orbiter hatch, inner panel, confined severance system.

high pressure is generated inside the lead tube and the outer steel tube. The combination begins to alter their initial unstable, semiellipsoidal shape to that of a more stable circle of equilibrium. The encircling XTA charge holder offers more resistance to the expanding tube than does the opposing panel, which has a geometric stress riser machined into it to control precisely the exact location at which the panel will break (sever). As the tube expands, it also wedges itself between the walls of the XTA charge holder and the remaining structure to prevent it from dislodging itself from its original location.

The severance device consists of a heat-resistant steel tube, 0.63 inch (1.6 centimeters) in outside diameter with a wall thickness of approximately 0.062 inch (1.6 millimeters), whose length is equal to the perimeter of the 30- by 48-inch (76- by 122-centimeter) panel to be severed. Inside the steel tube is inserted a lead tube with an outside diameter slightly smaller than the inside diameter of the steel tube. The hole in the lead tube is just large enough to receive the 37.5 grains per foot (8.0 grams per meter) MDC. This MDC is also silver-sheathed and the core explosive is also HNS. The MDC is placed inside the lead tube and the entire assembly is bent to conform to the curvature of the panel to be severed. Once this has been accomplished, the formed assembly is partially flattened to a thickness of 0.45 inch (11.4 millimeters). At this stage of fabrication, the assembly acquires its XTA identification. As with the outer fuselage panel severance system, the two ends of the MDC terminate in HNS boosters, each of which is initiated by a separate detonator. If only one detonator should fire, it will initiate the adjacent acceptor end booster and the MDC that terminates within it. In effect, the MDC is initiated from only one end due to the differential functioning time of two individual initiating systems.

The performance and safety margins are demonstrated in the same manner as described for the outer fuselage panel unconfined MDC severance system with one major exception: a steel tube that will expand, sever a test panel, and confine the high explosive energy of a single strand of MDC will rupture when two strands of the same size MDC are fired simultaneously. Conversely, a steel tube that is strong enough to contain the high explosive energy of dual MDCs (of the previous example) fired simultaneously is too stiff and will not expand enough to sever a test panel when a single MDC is fired. The ductility requirements for a steel tube to perform satisfactorily under these two extreme conditions are contradictory. The choice is to impose rigid quality assurance and control (see Chapter 8) on a single MDC, XTA severance system as was done on the orbiter.

## SHIELDED MILD DETONATING CORD (SMDC)

Shielded mild detonating cord (SMDC) assemblies are mild detonating cords (MDCs) contained within stainless steel tubing. The tubing with necessary end fittings for interconnection, as shown in Figure 4-8, is analogous to the

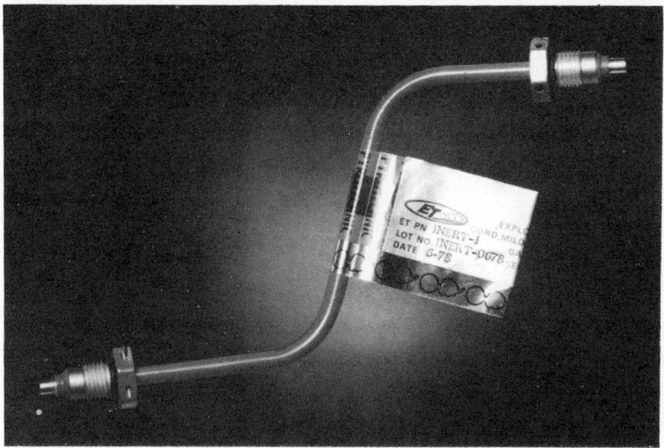


**FIG. 4-8.** Typical SMDC assembly.

tubing and fittings of conventional hydraulic and pneumatic lines in that they provide complete sealing for the transfer while confining the dynamic events (explosion) wholly within the plumbing. The SMDC plumbing technology is similar to that used in small-diameter hydraulic lines.

SMDC lines are used in installations where the presence of the crew or the nearness of sensitive materials or equipment makes it undesirable to use unconfined explosives. The principal application for SMDC lines is in stimuli-transfer networks of a multiplexed pyrotechnic system. The SMDC lines provide a safe, efficient means to interconnect related devices in a network to deliver activating stimuli to the functional devices of the system. The most frequent application for SMDC is in the stimulus-transfer network of an aircraft or manned spacecraft crew-escape system (see Chapter 7). Here the SMDC initiators are generally percussion primer type devices fired by spring-driven firing pins cocked and released by the hand action taken by the crewman (see Chapter 3). The SMDC lines carry the explosive stimuli from the initiators through various decision devices (mode selectors), manifolds, time-delays, and other implementing devices, delivering the energy to the numerous actuators, rocket catapults, severance systems, etc. within the system to execute the crew-escape event.

The most critical portion of the SMDC system is the interface between the donor (output) end of the SMDC and the acceptor (input) of the next device. To successfully transfer the explosive signal, great care is taken to provide donor and acceptor boosters on each side of the interface. Through work conducted in government laboratories (Naval Ordnance Labs), these boosters have been optimized into a standardized configuration used by all SMDC manufacturers.

The SMDC function is to transfer stimulus. The minimum energy level consistent with reliability is desired. When the explosive energy is low, the plumbing to confine the explosive events can be light and the overall SMDC system weight is likewise minimized. Figure 4-9 depicts the major components and geometry of a typical SMDC tip assembly. The stainless steel tube wall thickness is nominally 0.022 inch (0.55 millimeter) thick. The explosive core is generally HNS loaded at 2.5 grains per foot (49.4 milligrams per meter) drawn into an aluminum or silver sheath approximately 0.07 inch (1.8 millimeters) in diameter. When fired, the SMDC line expands approximately 15 percent. Bend radii as small as 0.75 inch (1.9 centimeters) are common. The explosive boosters on each end of an SMDC line composed of 1 grain (64.8 milligrams) of HNS explosive consolidated to 32,000 pounds per square inch (2,250 kilograms per square centimeter). SMDC line lengths are limited only by restrictions imposed by transporting, geometry of the configuration, and ease of installation and removal.

Because the transfer across the gaps between interfacing SMDC lines is the most critical concern for reliability in any SMDC network, the design of the booster tip has been standardized. The orientation of adjacent tips and gaps between them have been determined empirically by development testing for the most reliable detonation stimuli transfer. In brief, the experience shows:

1. Detonation transfer across gaps is accomplished by shock and fragmentation.

2. Orientation of the end tips has an influence on the gap distance across which the detonation can be transferred.

3. Reliability of energy transfer is directly proportional to the gap distance, except in some tip orientations there is a minimum gap allowable.

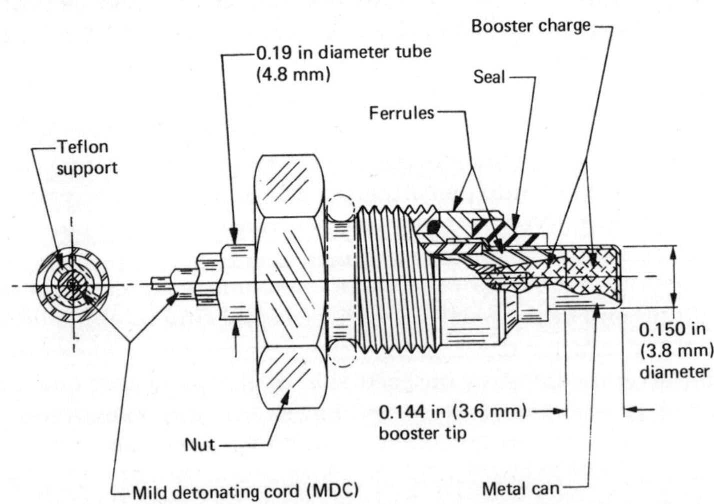


**FIG. 4-9.** Typical SMDC booster tip assembly.

**4.** A specific fragmentation pattern is emitted from the fired tip. This pattern, as shown in Figure 4-10, is sufficiently reproducable for use in establishing tip-to-tip relationships which assure the maximum fragmentation impingement.

It should be noted that the density of the fragmentation pattern decreases more rapidly from the SMDC tip in the radial pattern than in the axial pattern. This is due, as previously stated, to the fact that explosive energy is radiated perpendicularly from the surface of the explosive. From the cylindrical walls of the tip the energy rays are diverging, whereas from the flat end they remain almost parallel.

Of the many possible arrangements of interfacing SMDC tips, the orientations of Figure 4-11 (documented by test data) have produced the highest probability of reliable propagation across an air gap. The diagrams are numbered in the order of their functional dependability, the best being number one. The numbering sequence likewise reflects which tip orientations

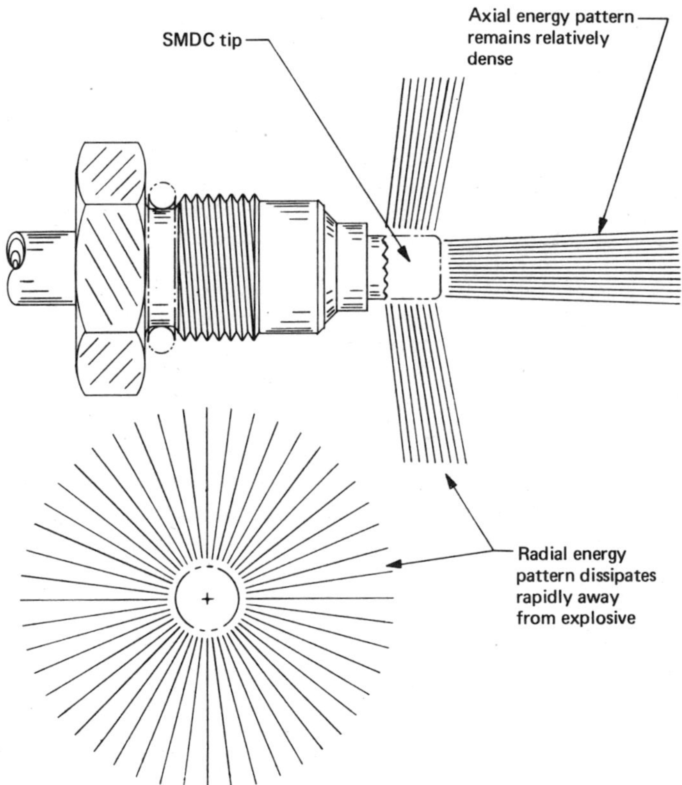


**FIG. 4-10.** SMDC booster tip fragmentation and energy dispersion patterns.

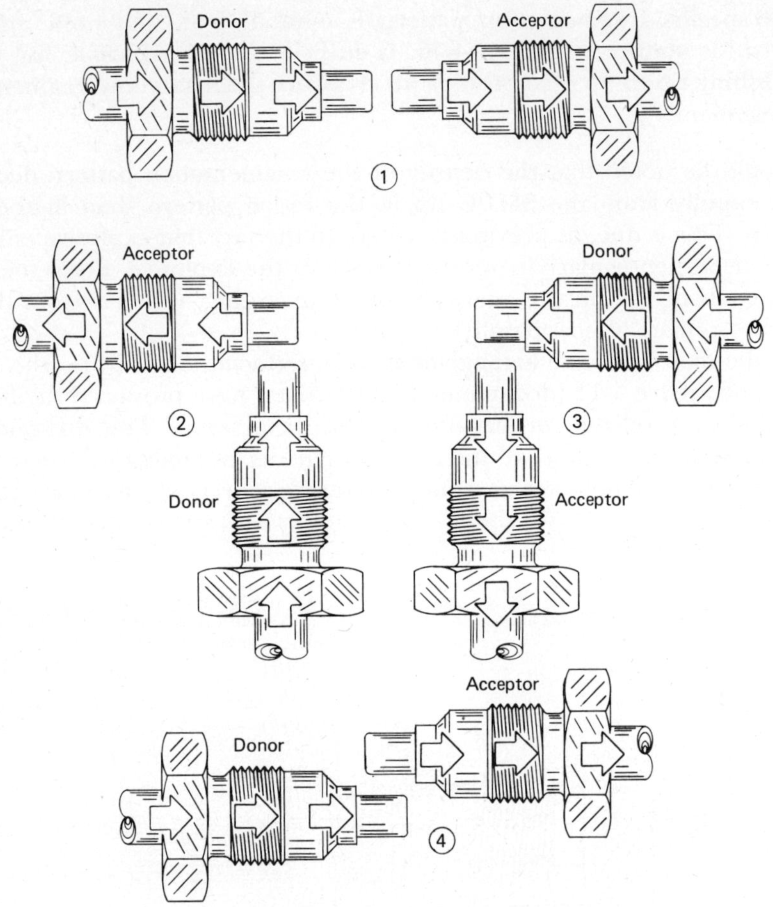


**FIG. 4-11.** Preferred SMDC tip-to-tip orientations.

are the least sensitive to an increase in tip-to-tip gap. Properly gapped, each of the orientations have demonstrated satisfactory reliability. The diagrams merely represent the order of preference for the idealized case.

The standard booster tip has been designed for high reliability of detonation transfer over gaps to 0.25 inch (6.4 millimeters). Surprisingly, the optimum gap is not "zero," i.e., tips in contact, since this does not give the required minimum distance for the fragmentation contribution to accelerate to its optimum momentum level. Gaps between 0.05 and 0.10 inch (1.28 and 2.56 millimeters) are best for maximum fragmentation effects. A nominal gap of 0.075 inch (1.92 millimeters) and reasonable manufacturing tolerances on tips and plumbing fittings make near optimum SMDC tip installation readily obtainable.

Where interfacing SMDC tips are in multiple arrays, the considerations of reliability become more complex. For example, where three SMDC tips are arranged in a tee arrangement having one tip as a donor and two as acceptors, there are a number of arrangements. Recalling the end-to-end is the first choice; it is obvious the donor and one of the acceptors will be in the end-to-end mode. This leaves two possible arrangements for the second acceptor: (1) to be initiated in the side-to-end mode by the donor, or (2) to be initiated in the side-to-end mode with the first acceptor acting as a parasite donor.

## CONFINED DETONATING CORD (CDC)

The mild detonating cord (MDC) of a confined detonating cord (CDC) is identical to that of an SMDC, namely 2.5 grains per foot (49.4 milligrams per meter) of HNS explosive. The donor and acceptor tips are also identical to the SMDCs, i.e., 1 grain (64.8 milligrams) of HNS. Figure 4-12 shows a typical CDC assembly. The confining covering over the silver or aluminum sheath are multiple layers of flexible braids and a waterproofing Teflon tube. The final result is a flexible detonating explosive cord, capable of containing all of the detonating products of combustion.

CDC has many applications such as transferring a detonating stimuli from a fixed structure across a hinge onto a movable door, or from an aircraft

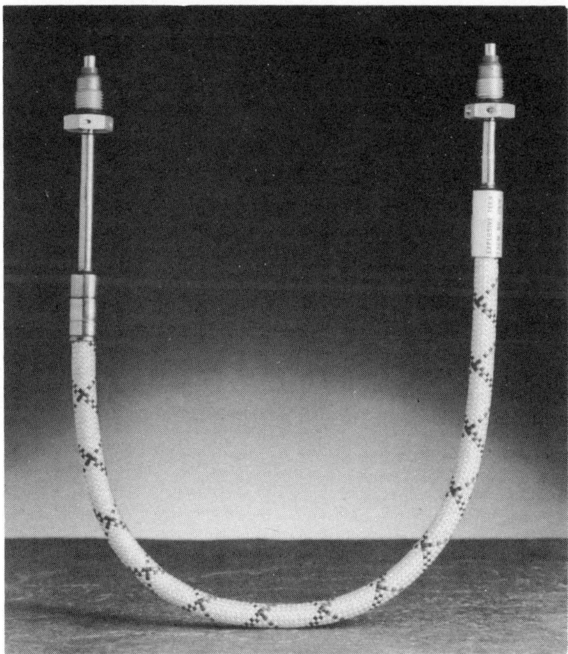

**FIG. 4-12.** Typical CDC assembly.

cockpit floor onto an ejection seat that will have to be adjusted to accommodate individual pilots.

CDC demands more care in handling and protection after installation than SMDC. This is due to the possiblity of the CDC being momentarily bent with too small a bend radius or of having been flexed too many times. Either extreme can cause permanent damage to the internal MDC without any apparent evidence on the braided cover. Abusive treatment of SMDC can be readily detected by observing the telltale sign of collapsed, dented, nicked, or scarfed metal tubing.

## DETONATING CORD

Detonating cord is a round, flexible cord containing a center core of high (or secondary) explosive, usually RDX or PETN. Unlike MDC, it has no metal sheath. The core is covered with various combinations of materials, such as textiles, waterproofing materials, asphalt, and plastics, so as to protect it from damage caused by physical abuses or exposure to extreme temperatures, water, oil, or other elements, and yet provide other essential features, such as tensile strength, flexibility, and safe handling characteristics. It must be remembered that detonating cord makes no attempt whatsoever to confine the detonating affect of the explosive core as does SMDC and CDC. In exterior appearance, detonating cord resembles CDC without donor or acceptor end fittings.

The purpose of detonating cord, as used in this book, is to initiate other explosive acceptors by the direct output blast of its detonating core. Without an intermediate booster, the core loading is understandably large, on the order of 25 to 50 grains per foot (5.3 to 10.6 grams per meter) with cord diameters of 0.16 and 0.20 inch (4 and 5 millimeters) respectively. The cord itself must be initiated by a blasting cap (see Chapter 3). A widely used detonating cord is marketed by the Ensign-Bickford Company under the trade name Primacord.

chapter 5

# Shaped Charges

Shaped charges are the only form of pyrotechnics that attempt not only to control the direction of the explosive energy but also to concentrate it. In many instances the energy intensity of an explosive can be increased more than a thousand times. Later in this chapter a comparison of the quantities of explosives required to perform a given task will be made with and without the technological precision of shaped charges.

## MONROE EFFECT

During the nineteenth century, miners learned that detonating sticks of dynamite that had one end arranged in a circle on the ground with their opposite ends tied together in the shape of a cone resulted in a much deeper hole being blown into the earth than if the same number of dynamite sticks were simply tied together with their axes parallel as shown in Figure 5-1. Although the miners did not understand the physical laws that had caused this phenomenon, they unknowingly had discovered the rudiments of shaped-charge pyrotechnology. The causation of this phenomenon is due primarily to the fact that energy forces of an explosive radiate predominately perpendicular from the surface of the explosive. Figure 5-2 illustrates how the energy forces of the three dynamite stick arrangements of Figure 5-1 are distributed by analyzing an enlarged section perpendicular to the target (the ground line) and through the center of the three configurations. Even though the quantity and distribution of energy from each individual dynamite stick is the same, the relative position of adjacent sticks can greatly influence the total amount of useful work accomplished. The angle between opposing sticks in the conical configuration causes the explosive energy forces radiating from the inside surface of the cone to intersect near the centerline. Here the horizontal components of these opposing energy forces collide head on and cancel each other. The vertical components, headed toward the target, are cumulative. Since configuration *a* has a far greater quantity and concentration of cumulative vertical vectorial energy forces contacting the target, the resultant hole is much deeper even though all three

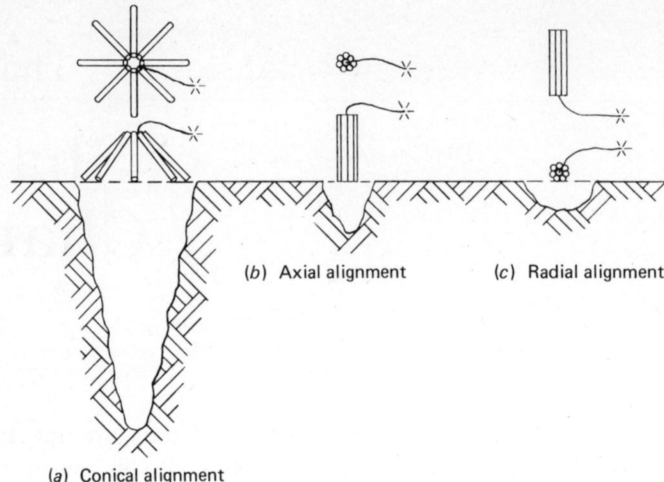


**FIG. 5-1.** Relative blast effects of eight dynamite sticks in three different arrangements.

configurations used the same number of dynamite sticks (8). Additional refinements of this basic discovery have enhanced even more the efficiency of shaped charges.

Shaped charges can be subdivided into three classifications: conical shaped charges (CSC); linear shaped charges (LSC); and flexible linear shaped charges (FLSC). The subdivisions have evolved as a result of their particular applications and will be discussed individually in this chapter. The following discussion of the principles of shaped-charge pyrotechnology are applicable to all three subdivisions. They were first analyzed and tested in 1888 by Charles E. Monroe. He observed that when the engraved lettering in a block of explosive was placed next to a metal plate and detonated, the image of the lettering was imparted (engraved) in the plate. In honor of his efforts the principle of the shaped charge is known as the Monroe Effect.

Figure 5-3 is a simplified illustration of a shaped charge in cross section

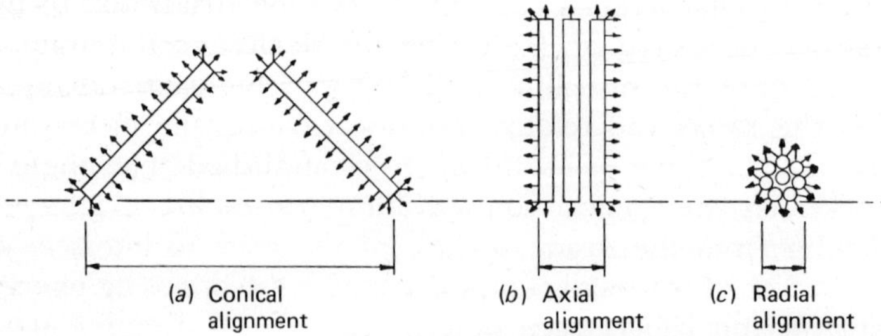


**FIG. 5-2.** Simplified dispersion of explosive energies of Figure 5-1. Arrows indicate area over which energy is distributed. (*Note*: Nearly half the total explosive energy of the conical alignment is utilized to blast hole in the ground.)

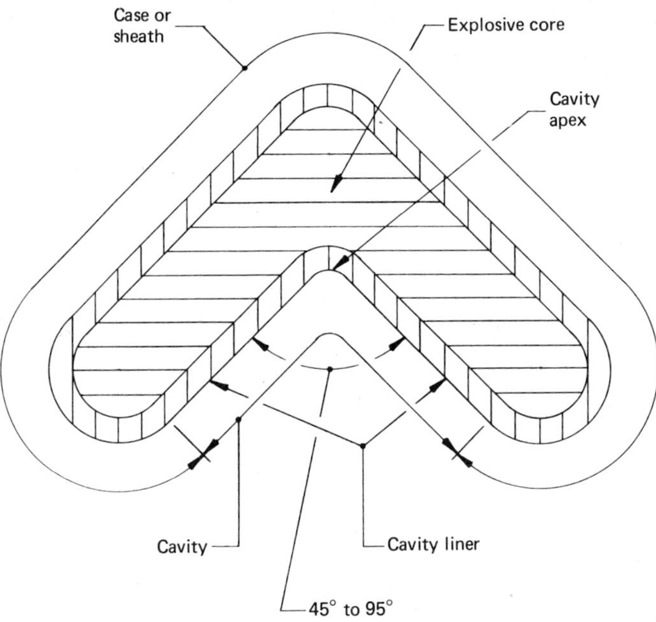


**FIG. 5-3.** Nomenclature of simplified shaped charge.

and identifies its major components. The housing or sheath and the cavity liner may be of one homogeneous material. Specific applications often dictate the two components be of different materials. Cavity liners are most often made of copper, lead, silver, and aluminum. Their primary purpose is to supply a source of heavy molecules that can be accelerated toward the target by the high pressure and shock waves generated by the high or secondary explosive core. RDX, PETN, HNS, and DIPAM (see Chapter 2) are the most often used core explosives. Upon impact with the target, these high-velocity molecules transfer tremendous amounts of kinetic energy to the target, causing it to deform. Comparative tests have shown that when all other parameters are equal, the performance of a shaped charge can be increased or decreased by increasing or decreasing the density of the liner material. Liner material preferences will be reviewed with specific shaped-charge applications. The included angle of the cavity liner varies between 45 and 90 degrees and it too is dependent upon the shaped-charge application. The interaction of the cavity liner's wall thickness, wall taper, and radius of the apex all affect performance.

The distribution of the explosive energy, as stated earlier, is essentially perpendicular from the surface of the explosive. Figure 5-4 details the energy distribution along the periphery of the simplified shaped charge. As illustrated here, even with the increased efficiency of the shaped charge, the majority of the explosive energy is wasted. Several unique shaped-charge

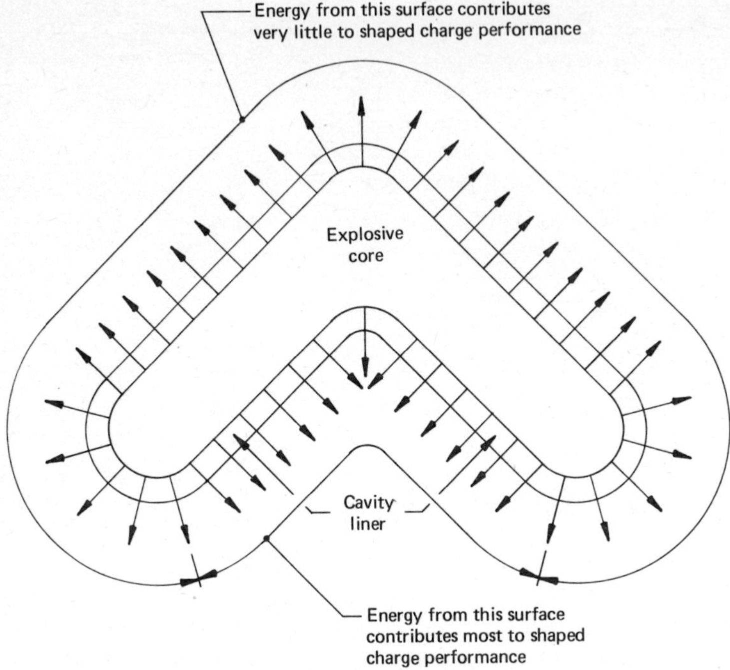


**FIG. 5-4.** Simplified shaped charge explosive energy distribution.

applications have made use of this apparently wasted energy and will be discussed later in this chapter.

The explosive phenomenon is generally described as the interaction of the detonation (rapid decomposition) products and cavity liner material emanating at high velocity from a shaped charge as the explosive detonates. The detonation releases large quantities of gas almost instantaneously under extreme pressure—as much as several millions of pounds per square inch or square centimeter. The shock waves emanating from the lower portion of a shaped charge converge at a point on the charge centerline and cause an extreme concentration of pressure along the axis of convergence as shown in Figure 5-5. These directed shock waves, together with the products of explosive decomposition and the metal molecules from the cavity liner, form the primary target cutting action—the jet.

Deformation of the target material begins within 1 microsecond (1 one-millionth of a second) after the passage of the detonation front. The shock waves produced by the expanding gases and the cavity liner material emanating from the lower portion of the shaped charge are converging, a jet of high velocity (in excess of 20,000 feet [6,100 meters] per second) cavity liner molecules is forming and penetration of the target is beginning. In Figure 5-6, the jet is fully developed and deformation of the target is well underway. The extent of this deformation is as follows: When a shaped charge is detonated on a metal target, the jet exerts an extremely concentrated force over a very small area. This force causes the metal to be pushed out of the way of the advancing jet by plastic flow. The resulting deformation is called "penetration." On a thin plate, however, the performance of the shaped charge depends not only on the cavity liner material and the intense, directed shock waves to erode the target, but also on the rapidly expanding gases to physically dislocate and fracture it. The shock waves, when reflected from the surface opposite the penetration, can also cause spalling (dislodged metal flakes) from that surface. The total effect is termed "cutting." Shaped charges consistently cut targets of greater thickness than they can penetrate.

The penetration action of a shaped charge is affected by a number of factors: the explosive used is of great importance; and while the depth of penetration is indicated to be more closely related to the detonating pressure than the rate of detonation, in general, the greater effect is produced by the explosive having the greater rate (velocity) of detonation. Very little effect is produced by explosives having rates of detonation of 15,000 feet (4,570 meters) per second or less.

Comparative tests with cavity liners of different metals give results that indicate, in general, the depth of penetration is greater with metal of higher density. However, liner ductility also plays a major role in penetration.

The standoff distance, or distance between the target and the base of the shaped charge required for maximum penetration effect, varies with the

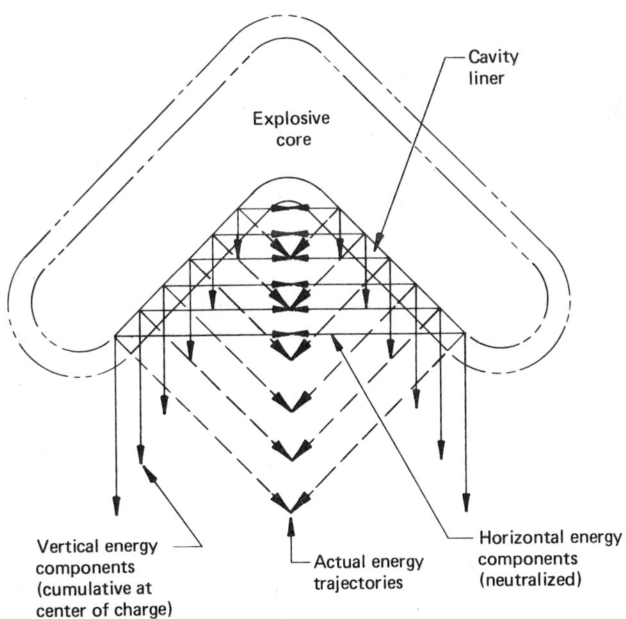


**FIG. 5-5.** Resolution of simplified shaped charge cavity energy.

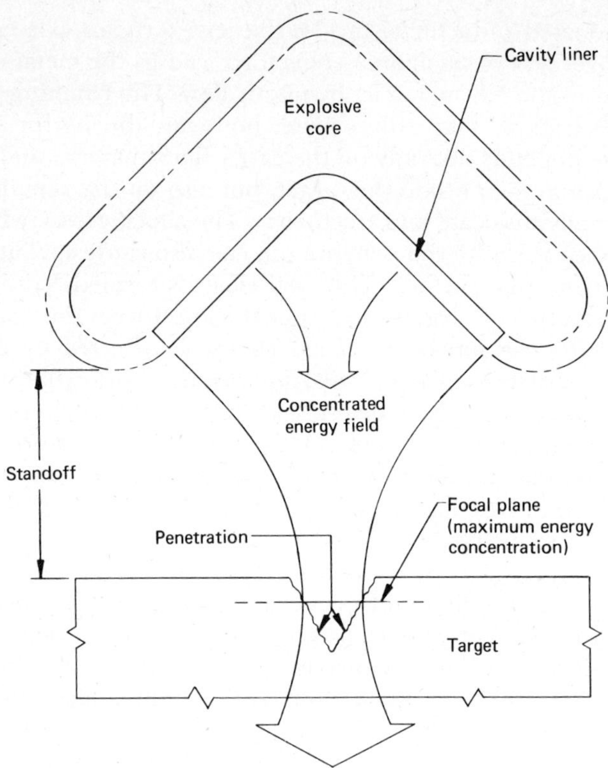


**FIG. 5-6.** Simplified shaped charge with optimum standoff for maximum target penetration.

metal used as a cavity liner. With a given liner, there is a given optimum standoff distance above and below which less penetration effect is obtained. As stated previously, the jet is the penetrating agent, and as standoff distance is increased, there is more time in which the jet can be extended. However, after a certain standoff distance, the jet has a tendency to break up both axially and radially.

Standard cutting data are usually derived under optimum conditions and usually tested against only one or two standard materials such as aluminum or steel. Data published by this empirical method can only be used as an approximation in selecting the proper size shaped charge to sever or penetrate a particular target. Other target parameters such as homogeneity, temper or hardness, direction of grain, and backup material, in conjunction with other shaped-charge parameters such as voids in the explosive, non-uniform compacting of explosive, and unsymmetrical geometry of the chevron, can affect the total performance of a particular shaped charge with a specific target. Only through rigorous manufacturing quality assurance and controls and thorough testing can an optimized, reliable shaped charge be configured (see Chapter 8).

## CONICAL SHAPED CHARGE (CSC)

A conical shaped charge is, as the name implies, a body of revolution rotated around the axis of symmetry. Table 5–1 illustrates a typical CSC and tabulates the geometric characteristics and performance capabilities for a family of

**Table 5-1.** Geometric, Weight, and Performance Characteristics of Conical Shaped Charges

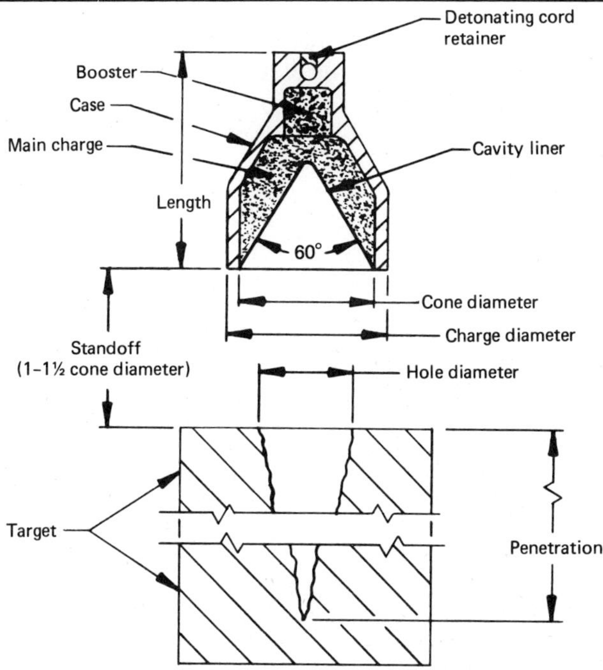


Copper cavity liner, RDX explosive

| Shaped charge number | Explosive weight, g | Gross weight, g | Approximate OD, in | Approximate overall length, in | Penetration, in* | Hole diameter, in* |
|---|---|---|---|---|---|---|
| 1 | 1.1 | 20 | 0.63 | 0.83 | 0.75 | 0.20 |
| 2 | 3.7 | 48 | 1.00 | 1.32 | 2.00 | 0.30 |
| 3 | 8.5 | 96 | 1.61 | 1.74 | 2.50 | 0.40 |
| 4 | 15.5 | 152 | 1.90 | 2.09 | 3.21 | 0.46 |
| 5 | 19.0 | 189 | 2.00 | 2.25 | 3.40 | 0.52 |
| 6 | 11.5 | 106 | 1.62 | 1.75 | 4.60 | 0.31 |
| 7 | 20.0 | 205 | 2.06 | 2.35 | 5.50 | 0.37 |
| 8 | 414.5 | 743 | 3.50 | 6.00 | 14.0 | 1.75 |

*Performance in mild steel

charges. The explosive material used in these CSCs is granular RDX (see Chapter 2) compressed into the case under pressure in excess of 15,000 pounds per square inch (1,055 kilograms per square centimeter). The tremendous amount of explosive energy released and focused by the CSC configuration is emphasized by the last entry in the table: Into a mild steel target a hole 1.75 inches (44.5 millimeters) in diameter with a penetration of 14 inches (356 millimeters). Users of these or slightly modified CSCs are the oceanographic industry for cable cutters, armed forces for demolition, construction contractors for drilling aids, and the steel industry for tapping open-hearth furnaces. The oil industry's application for perforating oil well casing is simplified in the illustration of Figure 5-7. By lowering a detector into a well casing, geologists are able to locate oil deposits in stratum at considerable distances from the casing itself. The problem of tapping into

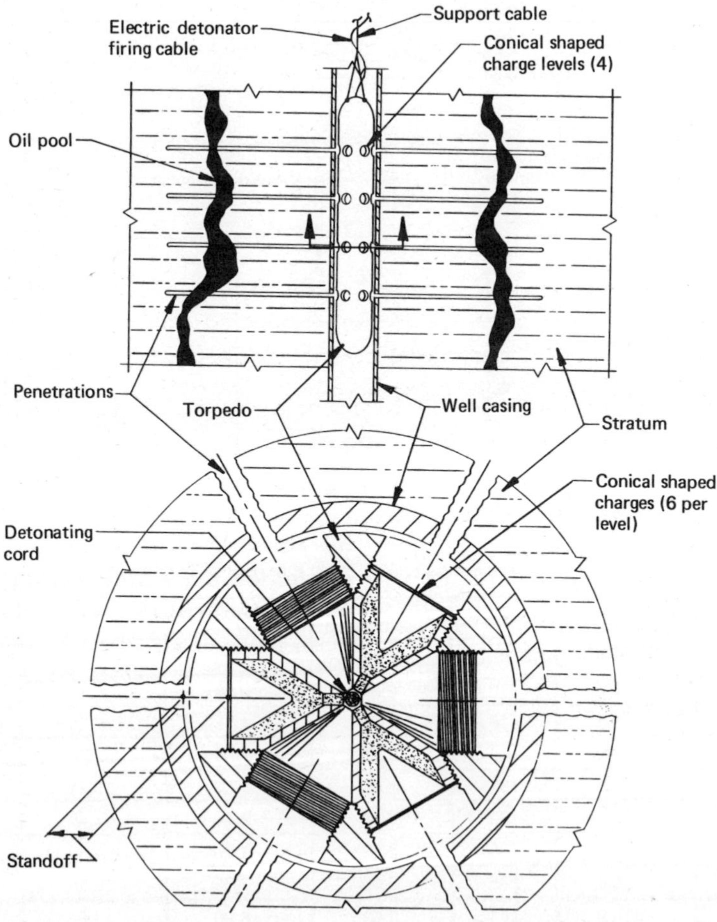


**FIG. 5-7.** Oil-well casing and stratum penetrators.

the adjacent oil pools without the added time and expense of drilling another well is resolved by inserting a "torpedo" into the casing and lowering it to the level of the oil pool(s). The torpedo consists of several levels of CDCs. The illustration depicts four levels with six CDCs per level. Torpedoes with twelve levels and twelve CDCs per level have been used. The torpedo is attached to a cable with a bridle at one end. An electric firing cable is entwined around the support cable and terminates at a detonator at the top of the torpedo. The detonator detonates a string of detonating cord (see Chapter 4) that traverses the length of the torpedo. As can be seen in the section, the detonating cord is located at the hub of the six radially oriented CSCs. The detonating cord simplifies the detonation of twenty-four CDCs with a single detonator.

When detonated, the CDCs penetrate not only the well casing but also the surrounding stratum for a considerable distance. Penetrations of several hundred inches (nearly a thousand centimeters) are not uncommon in Berea sandstone (a common oil-bearing stratum). Penetration is primarily dependent on the number of conical-shaped charges nestled into each level of the torpedo, the inside diameter of the well casing, thickness of the casing, casing material, number of concentric casings (up to four is not unusual), and the composition of the oil-bearing stratum. After the CSCs have penetrated the oil pool, the oil will immediately flow into the voids of the penetrations and through the holes in the casings. When inside the casing the oil is easily pumped to the surface.

Torpedoes have been built with a single CSC at each level. Obviously, a single charge designed to the full diameter of the torpedo will have greater penetration than multiple CSCs designed within the same diameter. To preclude the necessity of radial orientation of the torpedo with a single CSC at each level, a different radial orientation of each CSC is employed at each level.

The discovery of petroleum deposits under the sea created the need for many new techniques to recover oil from the new source. Problems connected with drilling platforms, special support vessels, and in many other areas were resolved by modifying techniques used in the standard land recovery methods. The task of laying the pipeline from the floor of the sea to the shore presented a unique problem in many areas, particularly in rock- and coral-encrusted coastal reefs and shoals. These areas quite often are shallow in depth, which precludes the use of deep-draft floating platforms from which to dig a trench for the pipeline to rest protected from the elements and possible entanglement with ships' anchors. In some cases the rock and coral formations are too hard to be economically removed with standard ditching or dredging equipment. One solution to the problem is the classical method of drilling bore-holes in the rock or coral formation, placing an explosive charge in each hole, and blasting to break or crush the dense formation. Needless to say, this is a slow and costly process.

The final solution came with the development of a conical shaped charge about the size of a small milk can. Instead of filling the CSCs with granular or solid explosives, and transporting them under rigid safety regulations to faraway places where they were needed, a liquid explosive was developed whose constituents can be shipped in separate containers by commercial transport to the using site. As can be seen in Figure 5-8, even when filled with the liquid explosive, more than half of the internal volume of the CSC is void. If placed in water the CSC would float inverted. This anomaly is overcome by placing the base of the CSC in an oversized box and filling the gap with concrete. CSC case segments are often molded plastic and the cavity liner is a deep drawn steel cone. The total assembly weighs approximately 40 to 50 pounds (18.14 to 22.68 kilograms). Handholds are provided in the box to facilitate carrying the CSC on land and maneuvering it into position under water. The stable liquid explosive ingredients are mixed and poured into the case through a hole in the top. The stopper serves a dual purpose—it is also the detonator. To the detonator is attached a length of detonating cord.

Figure 5-9 is a series of pictures of the insensitive liquid explosive chemicals and CSC cases being transported to the using site. There the liquid explosive constituents are mixed; cases are assembled and placed in handling boxes where concrete is poured around the base of the CSC. The liquid explosive is poured into the case and topped with a detonator and a short length of detonating cord (see Chapter 3). Barges are loaded with numerous

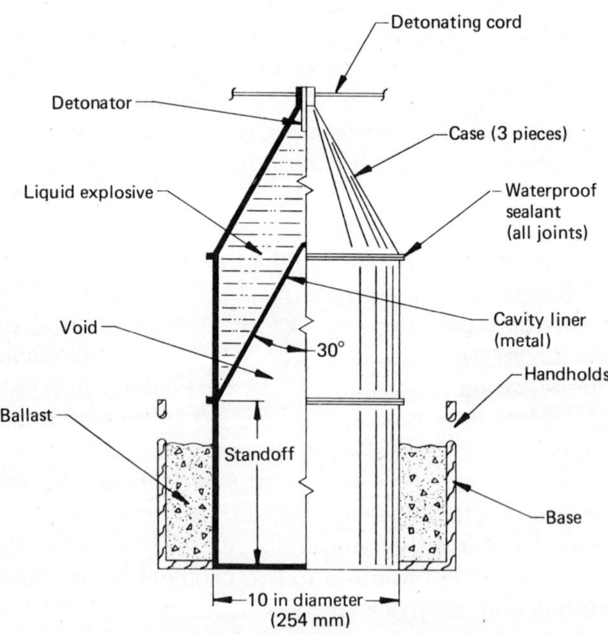


**FIG. 5-8.** Liquid explosive CSC used in trenching and dredging operations.

**FIG. 5-9.** Transporting, filling, positioning, and firing of CSCs for underwater oil pipeline operations. (*a*) Pallets loaded aboard air freighter. (*b*) Transporting pallets to using site. (*c*) Assembling CSCs. (*d*) Filling CSCs with liquid explosive. (*e*) Positioning CSCs along trenching line. (*f*) Detonating cord attached to CSC detonator. (*g*) Underwater trench being excavated. (*h*) Final positioning of oil pipe in trench through shoal.

CSCs for transporting to the underwater trenching site. At the trench site, up to several hundred CSCs are placed in rows about 4 feet (1.2 meters) apart on either side and along the centerline of the trench. The short lengths of detonating cord from each of the stationed CSCs are knotted to a longer detonating cord strung the length of the positioned CSCs. Attachment of an electric detonator to the end of the long detonating cords attached to the CSCs completes the trenching preparations. After detonating the CSCs, subsequent dredging is not usually required. The oil pipe is then placed in the trench and the job is complete. The following summary is typical of the underwater CSC trenching operations around the world:

**Mexico:** Isle DeLobo—Pipeline trench from offshore drilling platform to island storage facility.

*Trench specifications:* Width—Depth—Length—Feet (Meters) 14—6—2,300 (4.2—1.83—701) Operation completed in 4 days.

**Egypt:** El Alamain—Pipeline trench from mainland to offshore tanker loading facility.

*Trench specifications:* Width—Depth—Length—Feet (Meters) 16.5—6.5—1,-650 (5—2—503) Operation completed in 10 days.

**Iran:** Kharg Island—Pipeline trench from mainland to island storage facility.

*Trench specifications:* Width—Depth—Length—Feet (Meters) 40—8 to 14—3,050 (12.2—2.4 to 4.3—930) Operation completed in 25 days.

**Trucial States:** Jebel Dhanna—Pipeline trench from mainland to offshore tanker loading facility.

*Trench specifications:* Width—Depth—Length—Feet (Meters) 6—8—2,500 (1.8—2.4—762) Operation completed in 7 days.

**Alaska:** Cook Inlet—Pipeline trench from offshore platform to mainland.

*Trench specifications:* Width—Depth—Length—Feet (Meters) 4–7–2,500 (1.2–2.1–762) Operation completed in 4 days.

The same types of CSCs used in the above trenching operations can also be used in harbor and river dredging. Here the CSCs are placed in a checkerboard pattern, laced together with detonating cord, and detonated.

When solid rocket motors (SRMs) are launched, the capability must be provided to terminate the mission due to some malfunction of an onboard system, i.e., guidance, thermal control, etc. Unlike liquid propellant rockets, the SRMs cannot be shut down once they are ignited. If the SRM is of the type that has an open hole through its entire length, the propellant burns from the inside radially outwards over the entire length of the SRM. A simple method of terminating the thrust is to fire several conical shaped charges (by radio command) through the forward closeout dome of the SRM. The burning propellant will then exhaust through these forward holes, creating counter-thrust in the aft direction to neutralize the forward thrust created by the aft-firing rocket nozzle.

Some SRMs, however, don't burn radially their entire length. They burn from their aft end forward over the entire inside diameter of the rocket. Terminating the thrust of these SRMs necessitates reliable rocket igniters as well as conical-shaped charges. A dual-purpose CSC is shown in Figure 5-10 that includes a cylindrical exothermic pellet built into its base. The cylindrical geometry of the exothermic pellet allows the jet of the CSC to pass through its center after which it penetrates the forward dome of the SRM. The tail of the jet then ignites the pellet as it is pulled inside the dome of the SRM. Once inside, the burning pellet, 3,000°F (1,649°C) smears through the exposed SRM propellant—igniting it. Neutralizing thrust from the forward dome of SRMs is usually short-lived. The cylindrical walls will most often rupture a few seconds after the thrust termination CSCs have penetrated the SRM's dome due to internal overpressure.

## LINEAR SHAPED CHARGE (LSC)

In cross section, the linear shaped charge has many similarities to the conical-shaped charge with the main exception being the included angle of the cavity liner. Where the cavity angle of the CSC was shown to be approximately 60 degrees (see Table 5-1), the included angle of the LSC cavity liner is nearly 90 degrees. LSCs are generally fabricated in lengths up to 12 feet (3.66 meters) with explosive core loading up to 3,200 grains per foot (680 grams

**FIG. 5-10.** Cutaway of SRM thrust termination CSC with exothermic pellet.

per meter). Larger core loadings are generally 5 feet (1.5 meters) or less in length. LSCs of core loadings above 500 grains per foot (106 grams per meter) have the sheath material formed to the chevron shape and then the explosive core is incrementally added and compressed until the entire length of the sheath is filled. It is mandatory that the core compacting pressure be at least 8,000 pounds per square inch (562 kilograms per square centimeter). Otherwise, the detonation of the explosive core may stop or the shock wave front may slow down (this condition is called low-order detonation) without actually stopping. Experiments have shown (as stated earlier) that LSCs with detonation velocities of less than 15,000 feet (4,572 meters) per second have very little penetration effect on the target. LSCs with core loadings under 500 grains per foot (9.8 grams per meter) are made similarly to flexible linear shaped charges, i.e., round tubes of sheathing material are loaded incrementally, under pressure, with the desired explosive core material. After the ends of the tubes are sealed, they are repeatedly drawn through swaging dies to simultaneously stretch and form the chevron shape to the final explosive core loading. Table 5-2 depicts the geometry of a family of LSCs and tabulates their characteristics and performance capabilities.

Unlike conical shaped charges, where the depth of penetration of the jet from the cavity liner into the target determined the extent of its performance, a linear shaped charge's performance is determined by the maximum target thickness it can sever. The diagram in the table shows the depth of penetration is approximately half the thickness of the target severed. The unpenetrated half of the target is broken by the interaction of the high velocity, approximately 25,000 feet (7,620 meters) per second, shock waves that induce a crack to form at the bottom of the penetration and continue through the remaining thickness of the target. The shock waves dissipate as they travel away from the bottom of the penetration. The cracking mechanism is the collision of shock waves being reflected back toward the penetration from the opposite face of the target with additional shock waves traveling from the penetration. The collisions of the opposing shock waves induce high stresses in the target. When these stresses exceed the structural capability of the target material, it breaks. If, however, the remaining thickness of the target below the penetration is great enough to adequately weaken (attenuate) the reflected shock waves, the crack from the bottom of the penetration may not propagate all the way through the target. Then the target would not be severed.

The table also indicates the effectiveness of the heavier cavity liners in the overall performance of the LSCs. Compare the thicknesses of mild steel targets severed by copper sheathed LSCs and RDX explosive core loading of 300, 400, and 600 grains per foot with similarly loaded LSCs sheathed in aluminum.

A simple method of determining the optimum standoff of a specific LSC with a particular target material is shown in Figure 5-11. A target specimen

**Table 5-2.** Geometric, Weight, and Performance Characteristics of Linear Shaped Charges

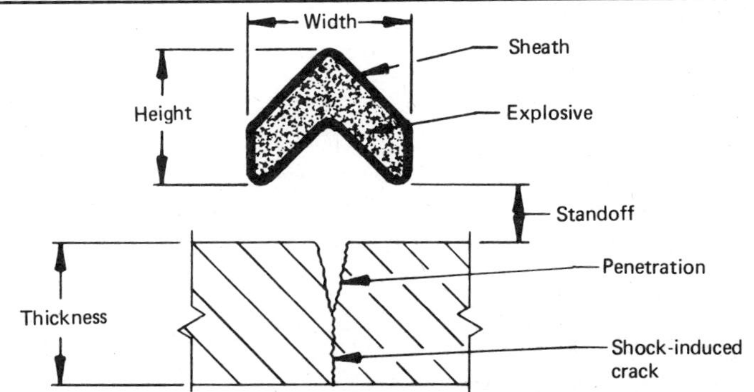


| Core load, grains/ft | Width, in | Height, in | Gross weight, lb/ft | Standoff, in | Thickness of target severed, in |
|---|---|---|---|---|---|
| Copper sheath, RDX explosive | | | | | |
| 100 | 0.28 | 0.25 | 0.07 | 0.20 | 0.32 |
| 150 | 0.35 | 0.31 | 0.20 | 0.20 | 0.40 |
| 200 | 0.45 | 0.38 | 0.21 | 0.25 | 0.50 |
| 300 | 0.45 | 0.40 | 0.22 | 0.37 | 0.62 |
| 400 | 0.48 | 0.53 | 0.31 | 0.37 | 0.75 |
| 600 | 0.68 | 0.58 | 0.51 | 0.60 | 1.00 |
| 1,400 | 1.15 | 0.94 | 1.25 | 0.75 | 1.37 |
| 2,000 | 1.15 | 1.04 | 1.31 | 0.75 | 1.75 |
| 3,200 | 1.43 | 1.23 | 1.66 | 1.00 | 2.12 |
| 4,400 | 1.81 | 1.41 | 2.50 | 1.25 | 2.50 |
| 10,500 | 2.56 | 1.78 | 4.30 | 2.00 | 4.00 |
| Aluminum sheath, RDX explosive | | | | | |
| 100 | 0.28 | 0.25 | 0.04 | 0.20 | 0.50* |
| 150 | 0.35 | 0.31 | 0.08 | 0.20 | 0.75* |
| 200 | 0.45 | 0.38 | 0.09 | 0.25 | 0.87* |
| 300 | 0.45 | 0.44 | 0.11 | 0.37 | 0.35 |
| 400 | 0.48 | 0.53 | 0.18 | 0.37 | 0.62 |
| 600 | 0.68 | 0.58 | 0.25 | 0.60 | 0.80 |

*Aluminum targets; all others mild steel

approximately 1 foot (0.3 meter) long and of greater thickness than the test LSC is anticipated to sever is selected. The test LSC should be 4 to 6 inches (10.2 to 15.3 centimeters) longer than the target. One end of the LSC is aligned with the end of and in contact with the target. The other end of the LSC is elevated such that the standoff distance at the end of the target is greater than the anticipated optimum. An electric detonator (blasting cap—

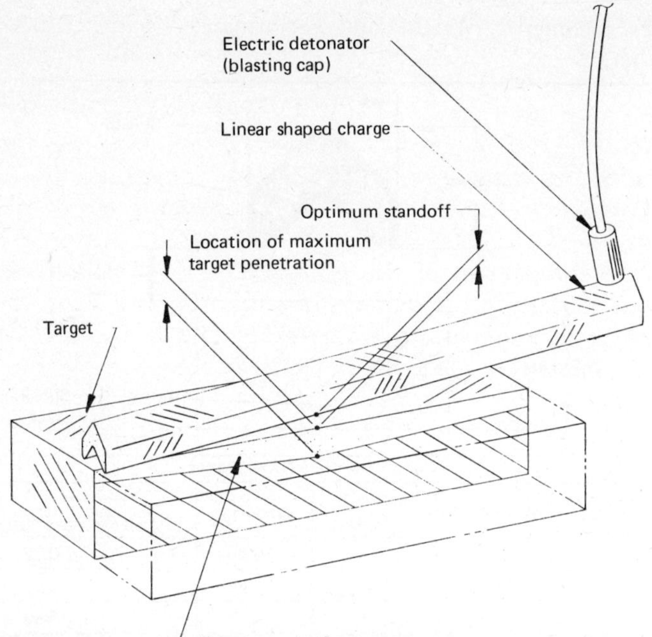


**FIG. 5-11.** Test set-up to determine optimum standoff of LSC.

see Chapter 3) is attached to the extended end of the LSC. Detonating the LSC away from the target allows the cutting jet (see Figure 5-6) to completely form before it contacts the target.

As the cutting jet of the detonating LSC propagates along the target, the depth of penetration will deepen as it nears the center of the target. Then it will begin to get shallow again as it approaches the opposite end of the target. The variance in the depth of penetration is caused by the variation in the standoff distance of the LSC across the length of the target. The standoff distance was deliberately made greater than optimum when the cutting jet first contacted the target; hence the penetration was shallow. Toward the center of the target the standoff distance lessened (approaching optimum) and the penetration grew deeper. Then as the cutting jet passed near the midpoint of the target and headed toward the far end, once again the penetration of the cutting jet lessened. The reason was that the standoff distance of the LSC became less than optimum. The location of the optimum standoff distance can be pinpointed as being at that location on the target where the LSC cutting jet penetration is deepest. Had the test results been that the penetration simply lessened from the initial target contact end all the way across to the opposite end, it would have meant that optimum standoff distance was never encountered and the test should be repeated with the elevated end of the linear shaped charge raised slightly higher. Repeating a test is seldom necessary because an approximation of the optimum standoff distance is readily available by referring to an LSC performance table such as Table 5–2. The elevated end of the LSC can be raised above the end of the target until the standoff distance is approximately twice that given in the table. If the explosive core loading of a test LSC should fall somewhere between two core loadings given in the table, the approximate standoff distance of the test LSC can be extrapolated.

A dramatic comparison of the effectiveness of LSC with conventional dynamite was evident in the demolition of the Central Ferry Bridge across the Snake River in southeastern Washington. The bridge was of steel girder construction, 1,450 feet (442 meters) long, made up of eight spans, and weighed nearly 200 tons. The bridge spanned a water basin that was to be deepened and would be beneath the surface of the new pool. Even though submerged, the bridge could be a hazard to boat navigation in the pool. A new, higher bridge had already been constructed within 50 feet (15.24 meters) of the bridge. Original estimates, by a demolition company using conventional dynamite, called for 225 pounds (102 kilograms) of dynamite per span or 1,800 pounds (816 kilograms) of dynamite to complete the job. Concern grew among the project managers about the effects of the blast from this huge quantity of dynamite might have on the nearby high bridge. This concern was alleviated when a second estimate was sought, and accepted, from another demolition company specializing in the use of precision linear shaped charges. Their estimate was to use only 3 pounds (1.36 kilograms) of explosives per span, or 24 pounds (10.89 kilograms) to demolish the entire bridge. Figure 5-12 shows a sequence of two of the eight spans being demolished simultaneously.

Most of the girders were severed with 2,000 grains per foot (425 grams per meter) LSCs. Figure 5-13 shows workmen installing LSCs on girders. A three-man crew completed the entire bridge demolition job in three working days. Only the two spans on opposite ends of the bridge were subsequently disassembled and hauled away. Since they fell on the shore or in shallow water, they were easily cut up with torches.

Such was not the case with the demolition of the bridge across the Mississippi River at Muscatine, Iowa. This bridge was a combination truss and suspension structure, 1,353 feet (412 meters) long, in three spans—each weighing approximately 25 tons. After the spans were dropped into the river they had to be removed. Each of the three spans was 451 feet (137.5 meters) long and would have to be removed from the bottom of the river in sections to facilitate handling. Since this was the main channel, the river was quite deep and cutting up the bridge underwater would be expensive, cumbersome, and time-consuming. The problem was resolved by simply cutting each span into six segments, as shown in Figure 5-14, with LSCs simultaneously while cutting them at the bridge pilings. Note the crane and barges

(a)

(b)

(c)

**FIG. 5-12.** Demolition of Central Ferry Bridge across Snake River, Washington. (*a*) Explosives installed . . . . Start the countdown! (*b*) Time: 0 plus 5 milliseconds! (*c*) Time: 0 plus 2 seconds!

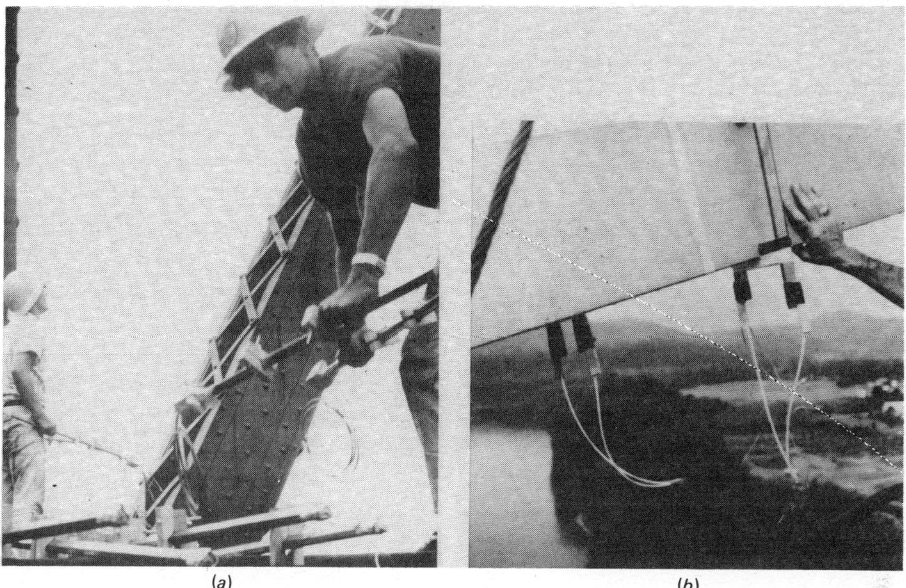

(a) (b)

**FIG. 5-13.** Preparing bridge girders for demolition. (*a*) Workmen ready to install LSC on bridge girders. (*b*) LSC installed on bridge girder.

beyond the bridge that moved in immediately and removed the bridge segments from the river in just 4 hours—a job that had been estimated to take more than 24 hours.

Like the conical shaped charges utilizing liquid explosives for underwater usage (see Figure 5-8), the cavity and the standoff distance of the linear shaped charge must remain void. This requires a hermetically sealed enclosure. Figure 5-15 presents LSCs specifically designed for underwater applications. Several LSC manufacturers stock hermetically sealed charges that will readily attach to a variety of standard structural steel members such as H beams, I beams, tees, zees, angles, and bars. Most demolition companies using LSCs can also assemble hermetically sealed charges for nonstandard structural members from components transported to and assembled at the demolition site.

## FLEXIBLE LINEAR SHAPED CHARGE (FLSC)

With the advent of linear shaped charges, it soon became apparent that there was a need to sever targets whose surfaces were contoured such as pipes, tubing, tanks, spheres, etc. of all diameters, thicknesses, and materials. Flexing a length of LSC without precise tooling can result in degradation of its performance. Even with close tolerance tooling, very little LSC with explosive core loadings in excess of 500 grains per foot (106 grams per meter) is flexed. The problem arises from the deformation of the optimum

(a)

(b)

(c)

**FIG. 5-14.** Demolition of high bridge across Mississippi River at Muscatine, Iowa.

(a) (b) (c)

**FIG. 5-15.** Hermetically sealed underwater LSC cutters. (*a*) Thick plate cutters. (*b*) Structural beam cutter. (*c*) Anchor chain cutter.

LSC chevron geometry. Without the aid of tooling to support the LSC while being flexed (bent), the two legs tend to straighten out or move toward each other. Either anomaly defeats the purpose of the optimum chevron geometry. Lead, aluminum, and silver are the predominant sheathing materials tolerant of severe forming operations. Through manufacturing experimentation, the LSC chevron shape (see Table 5-2) has been modified to be more compatible to flexing, hence the flexible linear shaped charge (FLSC). (Comparative performance tests of identical sheath, explosive, core loading, and target materials indicate slightly superior severance performance of the LSC geometry over that of the FLSC.) The sheathing material is often work-hardened while being formed, and residual stresses remain. Aluminum and silver can be stress-relieved after forming. However, since stress-relieving is a heat-treating process, there is a limit to its application due to the temperature limitations of the various explosive cores within. These may range from 250° to 500°F (121° to 260°C) for periods up to 24 hours, depending on the explosive used. The amount of stress-relieving required is also dependent on the severity of the bend. Cutting small diameter holes in thick domes of small radii imposes severe stresses in the sheath and may require several stress-relief operations during fabrication.

The vast majority of FLSC used today falls within explosive core loadings of 5 to 150 grains per foot (1 to 32 grams per meter). Like the smaller sizes of LSC, FLSC is fabricated by incrementally filling a tube of sheath material

**Table 5-3.** Geometric and Performance Characteristics of Flexible Linear Shaped Charges

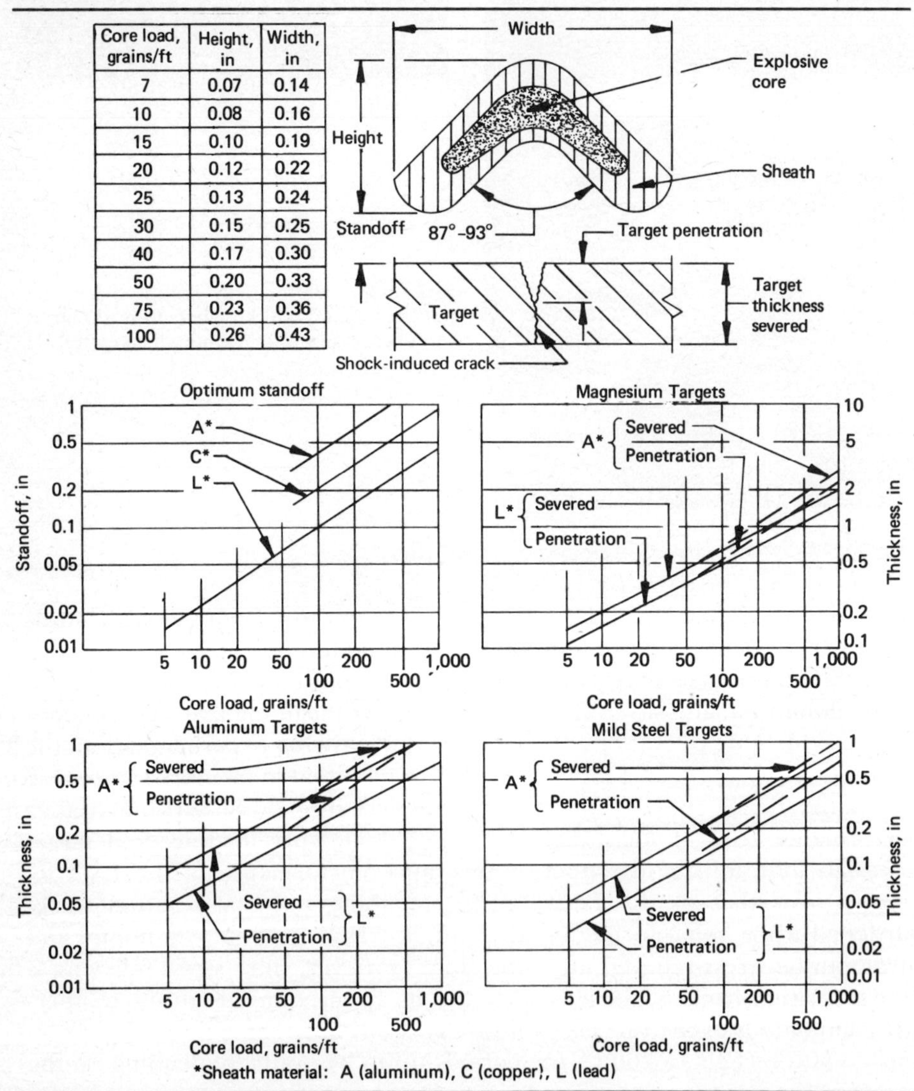

| Core load, grains/ft | Height, in | Width, in |
|---|---|---|
| 7 | 0.07 | 0.14 |
| 10 | 0.08 | 0.16 |
| 15 | 0.10 | 0.19 |
| 20 | 0.12 | 0.22 |
| 25 | 0.13 | 0.24 |
| 30 | 0.15 | 0.25 |
| 40 | 0.17 | 0.30 |
| 50 | 0.20 | 0.33 |
| 75 | 0.23 | 0.36 |
| 100 | 0.26 | 0.43 |

with compressings of the desired explosive, sealing the ends, and then drawing it through a combination of swaging and forming dies. Table 5-3 presents an FLSC cross section and tabulates the geometric and performance characteristics of a typical family.

The capability of FLSC expanding the usefulness of shaped-charge technology also imposes additional installation requirements. Unlike LSC, whose rigidity is a natural consequence of large cross sections and relatively short, straight lengths, they only require standoff support blocks (spacers) at either end (see Figure 5-13). The cross sections of FLSCs are much smaller, of greater lengths, and may contain numerous compound bends. To maintain proper standoff and alignment with the target, the FLSC requires continuous support along its entire length. Many materials have been used for supports, and the most popular are those that can be extruded and/or molded. These include polyethylene for ambient and low-temperature applications and silicone rubber or Teflon for ambient and high temperature. All these materials experience very little creep (deformation) while undergoing severe temperature excursions. Figure 5-16 is a representative sample of the types of FLSC continuous standoff spacers currently employed.

The continuous standoff spacer, while supporting the FLSC, must in turn be rigidly held against the target along a precise cutting line. This is accomplished by bonding the spacer inside a clip called an FLSC holder which is fastened to the target. These FLSC holders are often fabricated of

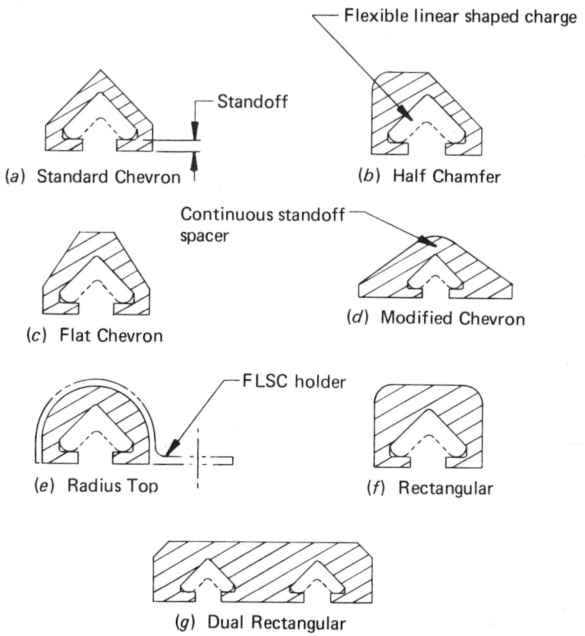


**FIG. 5-16.** Typical FLSC continuous standoff spacers.

fiberglass layups, formed sheet-metal, or lightweight machined fittings. Somewhere along the length of the holder will be an FLSC initiation block. This is usually a casting, forging, or machined fitting that supports a detonator in one of several possible orientations close to the FLSC, i.e., at the apex of the chevron (see Figure 5-11), tangent to either leg of the chevron or, if the FLSC extends beyond the edge of the target, the detonator may be nestled in the chevron cavity. The latter configuration is the most reliable since the shock-wave output of the detonator is in intimate contact with both legs of the FLSC chevron. To further enhance the detonation reliability of the severance system, multiple detonation blocks can be built into the FLSC holders. One of the most difficult problems associated with the long-time installation of FLSC systems is the assurance that moisture or other foreign matter does not collect between the FLSC cavity and the target. As explained earlier (see Figure 5-6), any interference with the focusing of the explosive energy greatly reduces the cutting performance of the FLSC. Many schemes have been tested with varying degrees of complexity and success. One such resolution used small hollow glass balls, two- to five-thousandths of an inch (0.05 to 0.13 millimeter) in diameter, called micro-balloons. They were bonded together (somewhat like popcorn balls) to the inside of the FLSC and standoff-spacer cavity. Although comparative performance test firings with and without the micro-balloons indicated a slight reduction in the FLSC performance with the balloons, they are quite effective in preventing moisture accumulation in the FLSC cavity and have experienced widespread use, especially in the aerospace industry. One objectionable side effect of micro-balloons is that they do create considerable quantities of glass dust when the FLSC is detonated. Unless some protection is provided to any nearby optical sensors, as in aerospace applications, the sensors could end up with a coating of powdered glass. Another method finding widespread use on earth application is to close out the opening at the base of the standoff spacer with a thin adhesive strip, 0.001 to 0.003 inch (0.025 to 0.050 millimeter), of aluminum foil. This concept would not work in aerospace applications because the air trapped inside the FLSC chevron/spacer cavity expands during ascent into space, causing the FLSC to be displaced away from the target. Since the initial position of the FLSC in relation to the target is its optimum standoff distance, the displacement results in degradation of the FLSC cutting performance.

An effective commercial application of FLSC is its use by fire fighters as forcible-entry hole cutters in walls, floors, and roofs of various types and combinations of construction. Table 5-4 portrays and tabulates the packages and hole sizes and geometries cut by the Jet-Axe, the trade name of a line of FLSC packaged cutters developed by Explosive Technology, a subsidiary of OEA. Each Jet-Axe package contains FLSC installed on the bottom of the waterproof housing in either a circular, square, or rectangular pattern. Attached to the housing is a loop of rope to suspend the Jet-Axe when used to cut holes in vertical walls. Each housing is provided with a punch-out

**Table 5-4.** Jet-Axe Forcible Entry Hole Cutters and Accessories

(*a*) A Family of Jet Axes

| Model | Package | Total package size/color | Total package weight | Cut profile | Cut config. | Total explosive | Typical applications |
|---|---|---|---|---|---|---|---|
| JA-I | □ | 32″ X 32″ X 4″ Yellow | 18.75 lb | □ | 24″ X 24″ | 3.5 oz | Mild steel plate to 0.4″ |
| JA-II | ○ | 20″ diam. X 6″ Yellow | 8.5 lb | ○ | 12″ diam. | 1.9 oz | Concrete or plywood to 3″ |
| JA-III | ▭ | 32″ X 44″ X 4″ Yellow | 27 lb | ▭ | 24″ X 36″ | 5.6 oz | Brick 2 courses |
| JA-IV | ○ | 26″ diam. X 7″ Red | 25.2 lb | ○ | 17″ diam. | 5.6 oz | Mild steel plate to 0.6″ Wood to 8″ |
| JA-V | ○ | 15″ diam. X 7″ Red | 11 lb | ○ | 10″ diam. | 3.4 oz | Concrete to 8″ Brick 3 courses |

Accessory Items

- ·Instructor's training manual
- ·Training film
- ·Magnetic hooks and masonry nails
- ·Firing cord extension (75′)

(*b*) Jet-Axe Characteristics

cover that facilitates access to a 50 foot (15.3 meters) length of CDC (see Chapter 4), and a CDC hand-held firing device (detonator). One end of the CDC is attached to a detonator mounted in the FLSC detonation block, the other end is free and sealed to protect the explosive from escaping and absorbing moisture. A typical CDC firing device consists of a pull-ring holding a firing pin against a compressed spring. When the pull-ring is pulled, the spring rams the firing pin against a percussion primer (see Chapter 3). After the free end of the CDC is installed into the end of the firing device, the pull-ring is pulled and the percussion primer detonates the CDC which in turn detonates the FLSC within the Jet-Axe housing. Figure

5-17 illustrates several ways the length of CDC enables a fireman to position himself a safe distance away from the backblast of the Jet-Axe after attaching it to a wall. He is now ready to install the firing device to the end of the CDC and pull the pull-ring. Behind the wall or door being cut by the Jet-Axe may be a room full of lethal gas or flame that could spew through the hole the instant it is cut. That potential danger is also alleviated when the fireman is a safe distance away from the Jet-Axe when it is detonated.

Figure 5-18 documents a few of the many instances the Jet-Axe has enabled firemen to greatly reduce the loss of life and property.

*(a)* An old wood wharf with several coatings of asphalt had a fire in the pilings and truss structure that hampered accessibility by the firemen. A Jet-Axe was placed on the wharf and fired, thus enabling the firemen to extinguish the fire promptly.

*(b)* Fire broke out in a Seattle second-floor dance studio. Firemen were prevented from attacking the fire via the stairs due to excessive smoke. When

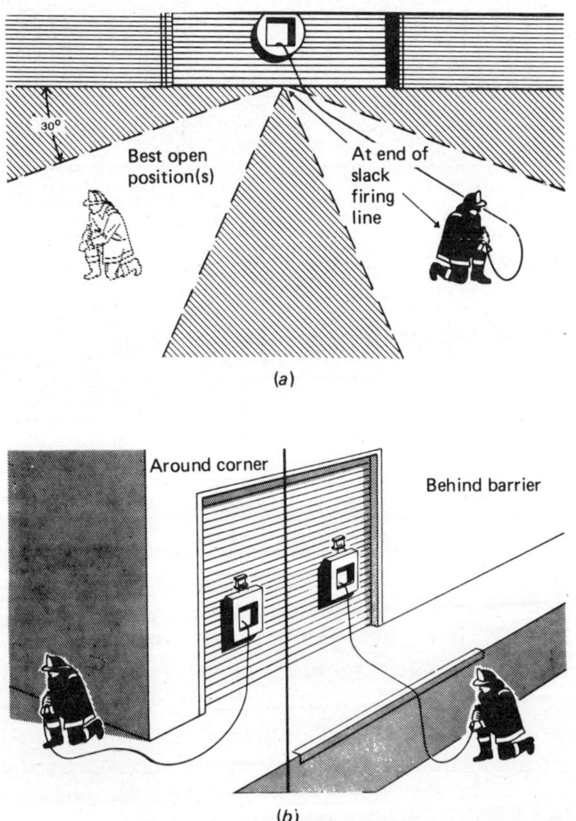


**FIG. 5-17.** Typical firing positions for wall-mounted Jet-Axe. *(a)* Firing position without cover available. *(b)* Firing position with cover available.

**FIG. 5-18.** Applications of Jet-Axe forceable entry FLSC cutters. (*a*) Wood and asphalt wharf. (*b*) Reinforced concrete. (*c*) Wood and composition. (*d*) Brick and stucco wall. (*e*) Girder-reinforced steel deck. (*f*) Partially reinforced steel deck.

a hole was attempted through the roof with a conventional hand axe, it was discovered the roof was composed of 2.5 inches (63.5 millimeters) of reinforced concrete. A Jet-Axe was placed on the roof and detonated, enabling the firemen to quell the fire in twenty minutes.

(*c*) The Zion, Illinois, fire department was summoned to a fire in a local restaurant. In addition to the smoke-filled first floor the basement was also completely charged with smoke since it was determined to be the point of origin of the fire. Entry could not be made into the rear of the building where the stairway to the basement was located. The fire began spreading upward and firefighting became extremely difficult. The basement ceiling (restaurant floor) was wood construction but had been remodeled three times, giving a thickness of three layers of flooring, tile and commercial pool carpeting. A Jet-Axe was placed on the floor and detonated. Through the hole belched forth superheated smoke and flames. Firemen inserted foam hoses through the Jet-Axe hole and were able to control the blaze in less than 20 minutes.

*(d)* When a fire broke out in the second floor of a downtown Evansville, Indiana, retail store, firemen discovered there were no windows on the second floor through which they could attack the blaze that was too intense to attack with hoses from the stairway. The fire had been burning unchecked for about 30 minutes because of this unaccessibility. Cutting through the stucco and brick facade of the building would have taken at least 30 minutes with conventional tools. The fire chief decided to have a Jet-Axe placed against the facade, over the raised lettering of the store's sign. Even though the lettering caused the Jet-Axe to be tilted slightly, it was fired successfully and the second floor was immediately flooded to put out the fire.

*(e)* and *(f)* The M.V. *Theresa Lee* is a refrigerated crab-processing vessel that was moored at Fisherman's Wharf in Seattle when a fire broke out in the forward hold containing knocked-down, tightly packed boxes of wax-coated cardboard cartons. The fire was located at the far end of the hold away from the entry hatch. It was decided to place a Jet-Axe on the overhead deck and cut a hole to gain access to the fire area. When detonated, the Jet-Axe had completely severed the half-inch-thick (12.7-millimeter) steel deck, but the panel did not fall into the hold as expected. Consultation with the ship's crew revealed that the forward hold is in the ice-breaking area of the ship's hull and therefore has additional reinforcing girders to prevent damage to the hull. It was two of these girders, backing up the deck plate, that prevented the severed plate from falling into the hold. The placement of a narrower Jet-Axe nearby, between two girders, provided the hole firemen used to insert foam hoses into the hold and smother the fire. Cleanup, replacement of the lost cargo, and repair of the minor damage delayed the ship's scheduled departure only a few days.

Because FLSC provides the greatest efficiency in the application of pyrotechnic energy, it has found widespread use in the aerospace industry. Its application to payload fairing separation systems has greatly increased the reliability of these systems because the number of pyrotechnic interfaces are reduced. A simplified sketch of the dynamics of a payload fairing being separated and jettisoned is depicted in Figure 5-19. The fairing is separated circumferentially around the base and simultaneously along its longitudinal plane of symmetry into two halves. After separation, the fairings are momentarily retained to the rocket at their bases by breakaway hinges until each has pivoted approximately 70 degrees outboard from their initial

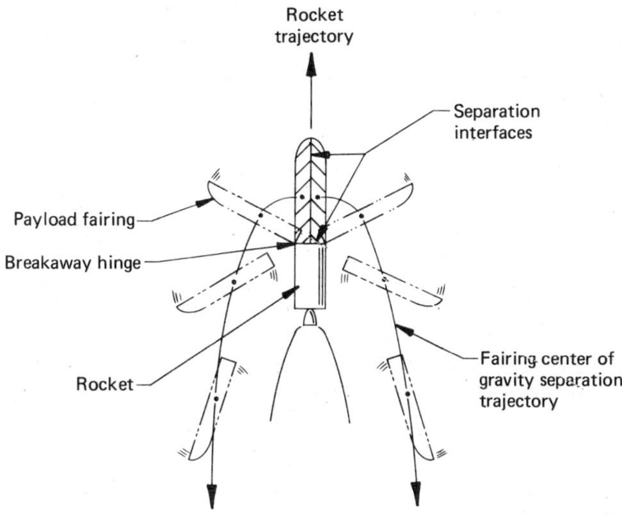


**FIG. 5-19.** Separation dynamics of rocket payload fairing.

positions. A comparison of two fairing separation systems that had been incorporated into this concept follows:

The first sytem used forty explosive bolts (see Chapter 6) to attach the two fairing halves together and to the rocket. Each bolt utilizes dual electro-explosive cartridges for redundancy—a total of 80 pyrotechnic devices to be initiated.

The second system employs splice plates and mechanical fasteners to hold the two fairing halves together and to the rocket. Onto these splice plates are mounted dual FLSCs (see Figure 5-16*g*) for redundancy. When detonated, the FLSCs sever the splice plates, separating the fairings. The total number of pyrotechnic components in this separation system is eighteen. This includes sixteen FLSC assemblies and two detonators. The reliability of the second system is much greater than the first because the number of pyro-technic interfaces is less.

The FLSC system is also lighter as the result of the interface loads between the fairings and the rocket are uniformly distributed along the separation planes, whereas in the bolted system the loads are concentrated at forty discrete locations. This necessitates stronger (and heavier) fairing edge members. It can be readily understood why the FLSC fairing separation system, offering higher reliability coupled with less weight, has widespread use in the aerospace industry.

A close scrutiny of Figure 5-19 discloses that without the aid of some auxiliary device to displace each fairing until their respective centers of gravity are displaced outboard of their hinge pivots, the two fairings may not be jettisoned either simultaneously or at all. This is due to the initial inboard location (in relation to the hinge) of the fairing's center of gravity. After the fairings are separated from the rocket and each other, the rocket's thrust is then imparted to the fairings through their respective hinges. The resultant opposing moments cause the two fairings to remain in their preseparated positions. Springs or pyrotechnic thrusters (see Chapter 6) are often used to overcome these forces and to initiate outboard rotation of the fairings until the thrust of the rocket can then supply the major jettison force.

One unique concept of jettisoning panels or fairings without the aid of auxiliary devices makes use of the generally wasted and sometimes detrimental FLSC energy that radiates from its upper surfaces (see Figure 5-4). Figure 5-20 illustrates an elementary application of such a system. The primary differences between this "sever and jettison" FLSC installation and the conventional "sever only" are the characteristics of the FLSC holders. Usually, FLSC holders are made of thin fiberglass or light metal that are completely obliterated by the FLSC energy radiated from its upper surfaces. If the FLSC holder were made strong and rigidly attached to the panel it was severing, then this upper surface energy would displace the FLSC holder, and hence the panel fastened to it, from their original positions. The upper

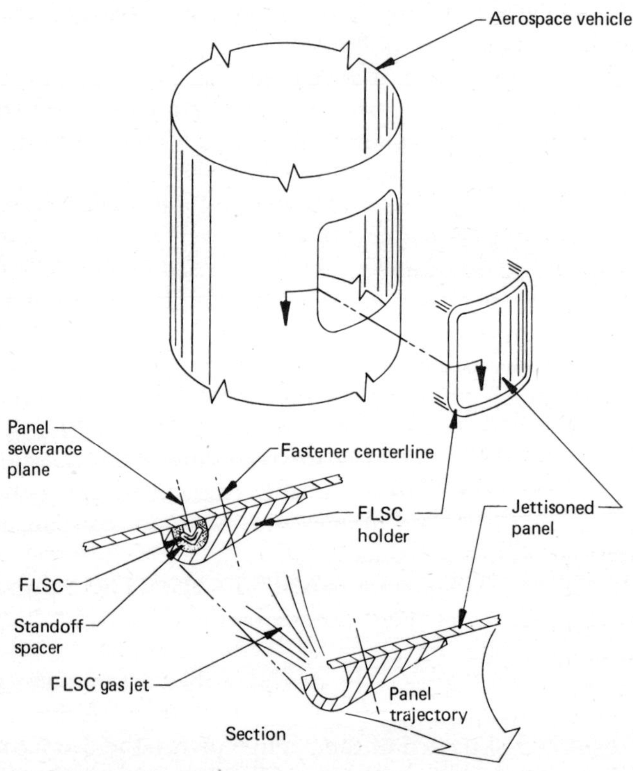


**FIG. 5-20.** FLSC panel severance and jettison system.

surface energy is also composed of high-velocity gases (similar to those focused to form the cutting jet from the FLSC cavity) whose trajectories are reversed by the presence of the rigid FLSC holder. The gases escape between the FLSC holder and the severed edge of the panel. The FLSC thrust-reversing jet augments the initial kinetic energy to help jettison the panel.

An instrumented test of a severed and jettisoned 20-inch (50.8-centimeter) square aluminum panel, 0.25 inch (6.35 millimeters) thick, weighing 14 pounds (6.35 kilograms), including FLSC holder and severed with 25 grains per foot (5.3 grams per meter) RDX FLSC with aluminum sheath, attained a velocity of 64.5 feet (19.7 meters) per second, measured at a distance of 5 feet (1.5 meters) from its initial position, with less than 5 degrees of angular rotation. Analysis of the instrumentation data indicated acceleration occurred over a distance of 0.4 inch (10.2 millimeters) with an average force of 7,130 pounds (3,234 kilograms) over and above that required to overcome air resistance. The perimeter of the panel, along the severed edge, was 75 inches (22.9 centimeters). A simple calculation reveals an acceleration force of 95 pounds per linear inch (17 kilograms per linear centimeter) of FLSC. If the panel had been jettisoned in the vacuum of space, the velocity would have been considerably greater.

Often the requirement is to sever and jettison whole cylindrical or conical fairings such as that shown in Figure 5-21. The entire fairing adapter, between the two rocket stages, must be severed and jettisoned to permit ignition of the second-stage solid rocket motor (SRM) and to enable its nozzle to gimbal (immediately, if necessary) beyond the confines of the fairing to provide SRM thrust vector control as shown in Figure 5-21. The circumferencial FLSC installation at the top and bottom of the fairing is identical to that for the panel severance and jettison of Figure 5-20. The installation of two vertical side-by-side strings of FLSC, no matter how close together they are placed, results in a sliver of fairing that can damage the SRM nozzle or impair the displacement of adjacent fairings. The problem is resolved by machining matching, tapered slots in the vertical FLSC holders of adjacent fairings, similar to a piano hinge interface. The staggered slots overlap a single vertical strand of FLSC and its standoff spacer, both of which are obliterated upon FLSC detonation. The tapers in the slots are tolerant of unmatched jettison trajectories of adjacent fairings. To prohibit water and other foreign matter from entering the FLSC cavity via the gap between the slots of matching FLSC holders, the gaps are filled with room-temperature vulcanizing (RTV) silicone rubber. The RTV offers no resistance to the jettisoning fairings.

Other applications of FLSC thrust reversing, severance, and jettisoning systems demonstrate the versatility of this sytem. Data packages, location aids (transponders, ocean dye markers, flashing lights, etc.), flight test instrumentation packages, and other payloads can be attached to jettisonable panels and subsequently recovered by parachute. Similar panels may have different payloads for different missions. To complicate the requirement even more,

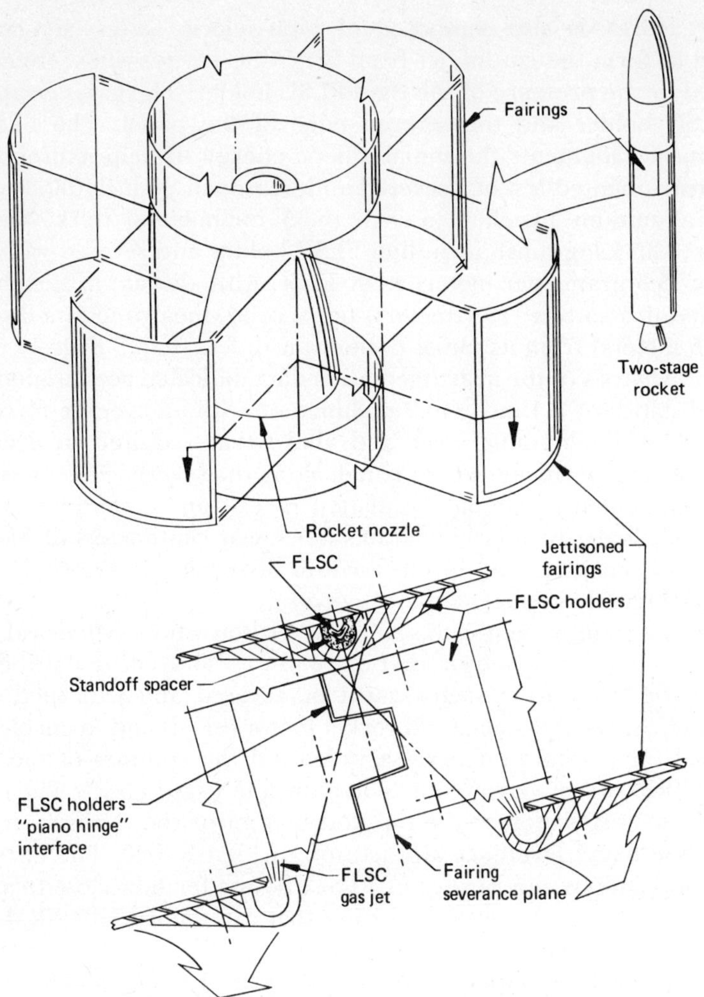


**FIG. 5-21.** FLSC multi-fairing severance and jettison system.

the center of gravity of the combined payload/panel may not coincide with the centroid of the FLSC pattern. This eccentricity would cause the payload/panel to tumble while being jettisoned, which could be detrimental to the deployment of the parachute recovery system. The resolution is to interrupt (remove) the FLSC holder (not the standoff spacer) at the proper location(s) to cause the centroid of the now intermittent thrust-reversing forces to align with the payload/panel center of gravity. This modification interrupts only the payload/panel jettison forces while maintaining continuity of the FLSC severance jet around the perimeter of the panel. The removed portion of the rigid FLSC holder is replaced with a lightweight holder that too is obliterated when the FLSC is detonated. The lightweight FLSC holder is necessary to maintain continuous support of the standoff spacer for reliable panel severance while at the same time contributing nothing to the payload/ panel jettison forces.

chapter 6

# Cartridge-Actuated Devices

Cartridge-actuated devices are the workhorses of the pyrotechnics industry. Their initiation stimuli may be any of the various types described in Chapter 3. The main pyrotechnic output charge may be either a secondary (or high) explosive (see Chapter 2), in which case the cartridge is called a "detonator cartridge," or it may be a deflagrating propellant charge (also described in Chapter 2) that produces high-pressure gas in sufficient quantity to perform useful work and hence is known as a "pressure cartridge." The actual amount of gas molecules produced is relatively small. The high pressure generated is due primarily to the high temperature that accompanies the decomposition of the deflagrating pyrotechnic propellant. This means the devices utilizing pyrotechnic pressure cartridges as power sources must perform their work quickly, otherwise the efficiency of the cartridge will be reduced by the heat of the decomposition being transferred from the hot gases to the surrounding chamber.

An example of an actual test demonstrates this phenomenon as follows: A particular pressure cartridge will produce pressure of approximately 300 pounds per square inch (21.1 kilograms per square centimeter) in a volume of 1 cubic foot (28.32 liters) in approximately 7 milliseconds after the cartridge is initiated. After 5 minutes or so have elapsed, enough heat will have transferred from the gas to the surrounding pressure vessel to cause the residual pressure to drop to approximately 29 pounds per square inch (2 kilograms per square centimeter). From this demonstration it can be seen that pressure cartridges have their greatest efficiency when the high pressures they produce can be utilized in short periods of time.

Because the pyrotechnic cartridge is separable from the inert device, this capability allows the inert functioning component, i.e., thruster, frangible bolt, guillotine, etc., to be installed without the added safety requirements necessary when installing a component containing integral pyrotechnics. In some instances the inert components of a pyrotechnic system may be installed months ahead of their pyrotechnic cartridge(s). Once the pyrotechnics have been installed, special precautionary procedures become effective, e.g., strict control of garments that could generate static electricity, strict electrical circuit control before turning power on the vehicle, and occasionally limiting the number of personnel that can be working on, or near, the vehicle at any one time.

Pyrotechnic cartridge-actuated devices have several distinct advantages over other types of power systems, such as hydraulics, pneumatics, electrical, etc. Their main advantage is being extremely light weight. This is due to the relative higher operating pressure of pyrotechnics versus hydraulic or pneumatics; namely 10,000 to 12,000 pounds per square inch (703 to 844 kilograms per square centimeter) versus 2,000 to 4,000 pounds per square inch (141 to 282 kilograms per square centimeter) respectively. All these components are generally made of steel or aluminum that permit relatively thin cylinder walls. To perform equal amounts of work, hydraulic or pneumatic actuators would require pistons with two to six times the area of a comparable pyrotechnic actuator and hence would be considerably heavier too. Additional system weight saving is realized when hydraulic or pneumatic tubing and pumps (or compressors) are replaced with a pair of pyrotechnic firing wires.

Another distinct advantage of cartridge-actuated devices is their simultaneity of actuation for multiple devices. This feature becomes extremely important when two or more attachments must be released at the same time, such as aircraft fuel tanks, multistage booster rockets, and interplanetary spacecraft. Examples similar to these and other applications will be reviewed throughout the remainder of this chapter.

## SEPARATION/RELEASE DEVICES

As the name implies, these devices are pyrotechnic cartridge-actuated separation or release devices that may structurally attach two adjacent components together and when initiated separate or release them. A more complex device may utilize a pyrotechnic device somewhat as a toggle stage, which upon initiation allows a mechanical device to perform the actual separation (or releasing) function.

Separation/release devices are available in two primary groups: frangible and nonfrangible devices. Within each of these two groups are various configurations such as bolts, nuts, and links, that will be individually addressed in this chapter.

### Frangible Devices

These devices are the most reliable of all pyrotechnic separation devices because they are, by far, the ultimate in simplicity of design. They are load-carrying structural members that are fractured by the explosive output of their pyrotechnic cartridges. In the selection of any frangible device, consideration must be given to the comparatively high pyrotechnic and mechanical shock imparted to the adjacent structure near the device. Avionics, thermal protection systems, and even structure itself have been damaged when the aforementioned concerns were ignored, especially in high-load-restraining applications. Other types of pyrotechnic separation devices, discussed later in this chapter, have either alleviated or eliminated this shock in other high-load applications.

**Frangible Nuts** The simplest of all frangible devices are frangible nuts. Figure 6-1 illustrates one of these devices that consists only of a conventional structural nut that has been modified to accommodate redundant, conventional output detonator cartridges (see Figure 3-12). One of these nuts was used to attach and release each of the four legs of the launch escape tower from the Apollo command module. In the event of an emergency occurring during the early launch phase of the Apollo spacecraft that would endanger

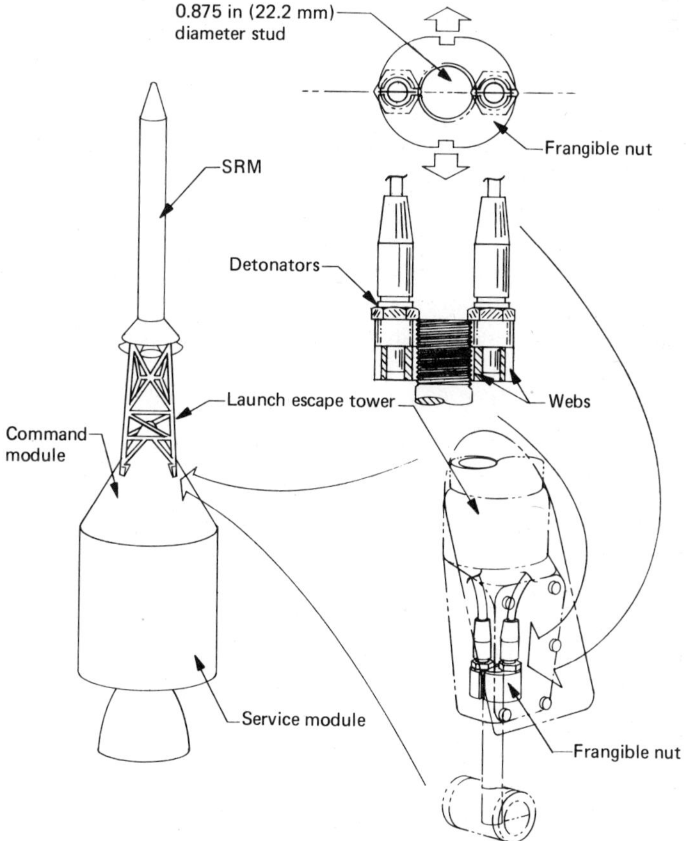


**FIG. 6-1.** Apollo launch escape tower separation frangible nut.

the lives of the three astronauts inside the command module, a launch escape solid rocket motor (SRM), attached to the launch escape tower, is initiated simultaneously with the releasing of the command module from atop its service module. The SRM pulls the command module some 5,000 feet (1,524 meters) above the service module, at which time the eight detonators in the four frangible nuts are initiated, thereby releasing the tower from the command module. This sequence also initiates the parachute recovery system that eventually safely lands the command module and its three astronauts back on earth. All the fragments and blast effects of the detonators are contained within the four tower legs.

Frangible nuts are based upon the principle of minimizing the amount of material required to retain together two integral halves of a nut when loaded to a predetermined maximum tensile load. Radial loads are imposed on each nut half, when a tensile load is imparted by a mating stud, along the bevels of the nut/stud interfacing threads. Neglecting the effects of interfacing thread surface roughness and friction, the radial load imparted to the nut is theoretically equal to the tensile load multiplied by the tangent of the interfacing thread flank angle. The Apollo frangible nut utilizes the standard military thread with 30-degree flank angles whose tangent is 0.577. This means the combined static tensile strength of the four webs holding the two halves of the frangible nut together will have to be capable of restraining slightly more than half the maximum expected tensile load applied by the stud. Naturally, this simple calculation is only the preliminary starting point in the design of a frangible nut. The final configuration is arrived at after several test specimens, with slightly thinner and thicker webs, have been statically loaded to ascertain the optimum web thickness. The thickness of the webs to be subsequently broken by the initiated detonators has a direct bearing on the minimum amount of explosive required to reliably break them. The quantity of explosive in two detonators for each of the four frangible nuts has a direct bearing on the blast containment strength requirements of each of the tower legs. If all the aforementioned optimizations are ignored, then the weight saving afforded by the simple frangible nut is negated by the necessity of a stronger (and heavier) blast container.

The frangible nut material is vacuum-melt, low-carbon steel, and when installed in a tensile test fixture and pulled until either the frangible nut or stud fails, the failure consistently occurs in the threaded portion of the studs. The thin webs holding the two halves of the frangible nut together are broken when one or both of the detonator cartridges are initiated. When only one detonator fires, it breaks the adjacent webs and spreads the two halves of the nut apart, causing the two webs on the opposite end to also break due to the fulcrum effect of the spreading nut halves. Even when both detonators are fired simultaneously, one detonator always preceeds the other due to differences in firing circuits and detonator characteristics. Instrumented tests have shown the nuts to be completely separated within 3 to 4 milliseconds after the onset of current flow in the bridgewire of the detonator.

The space shuttle (see Figure 7-1) utilizes three different diameter frangible nuts at three different element interfaces. Figure 6-2*a* illustrates these devices and their applications are as follows: Three 0.75-inch (19-millimeter) diameter frangible nuts are used at each of two orbiter/external tank liquid oxygen and liquid hydrogen umbilical disconnect interfaces. These nuts can sustain ultimate loads of 96,000 pounds (43,500 kilograms) and are similar to the Apollo launch escape tower frangible nuts previously discussed. Two of the 2.5-inch (63.5-millimeter) diameter frangible nuts provide the aft structural attachments and release devices of the orbiter from the external tank. Each nut is capable of sustaining ultimate static loads in excess of 600,000 pounds (272,000 kilograms). Four of the 3.5-inch (88.9-millimeter) diameter frangible nuts attach (and release) each of the two solid rocket boosters (SRBs) to the launch platform and are initiated simultaneously with the ignition of the SRBs. These giant frangible nuts are capable of ultimate loads in excess of 1,000,000 pounds (453,600 kilograms) each. All three space shuttle frangible nuts are made of Inconel 718 steel, which has extreme thermal environmental capabilities. All three are also broken by redundant detonators, while the two larger nuts also require dual detonator boosters. Figure 6-2*b* pictures the three frangible nuts together.

During the development of the two larger frangible nuts, a unique technique was employed to minimize the amount of detonator booster explosive required to break them. A reduction in the flank angle of the nut/stud interfacing threads was sure to reduce the magnitude of the induced radial loads. Engineers had to be careful with this approach because a coefficient of friction in excess of 0.20 at the thread interfaces could require a larger detonator to overcome than is required to break the frangible nut webs alone. To put it another way, if the thread flank angle were zero degrees, then there would be no radial load induced into the halves of the frangible nut when a tensile load is applied to the stud. This analysis holds true even when a flank angle up to approximately 5 degrees is used and is due to frictional resistance between the stud and frangible nut sliding, threaded interfaces. The optimum thread flank angle has been determined to be that of the standard 7-degree buttress thread when nominal shop machining practices and dry-lubrication processes are employed. When compared to the nominal 30-degree thread flank angle, the buttress thread's 7 degrees reduces the theoretical radial loads from the aforementioned 0.577 to 0.122 of the stud tensile load—a reduction of nearly 80 percent.

**Frangible Bolts** The main advantage of frangible bolts over frangible nuts is that the former do not require such massive blast and fragment containers as do the latter. As we shall see later in this chapter, blast and fragmentation can be virtually eliminated in the frangible bolt and yet fracturing of the primary load-carrying structural member, to effect a separation, can be accomplished.

The space shuttle orbiter has a unique requirement for a separation device

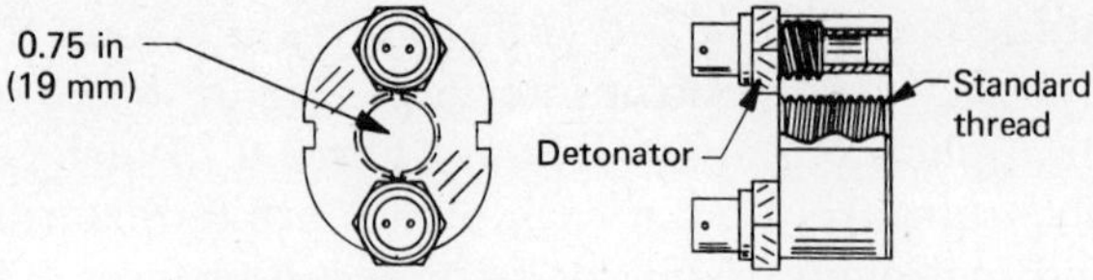


Orbiter/External Tank Umbilical Frangible Nut

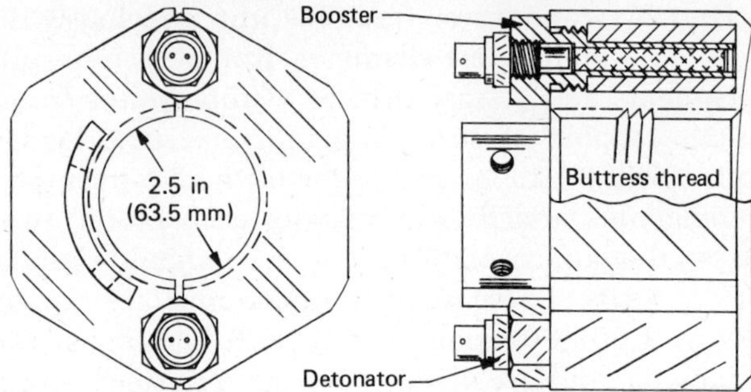


Orbiter/External Tank Aft Attach Frangible Nut

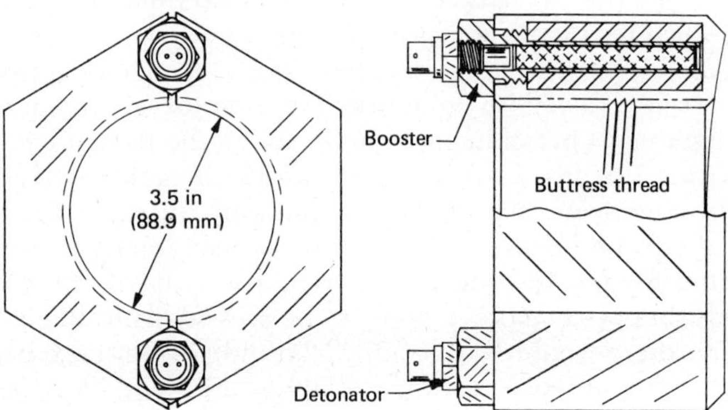


Solid Rocket Booster/Launch Platform Frangible Nut

(*a*)

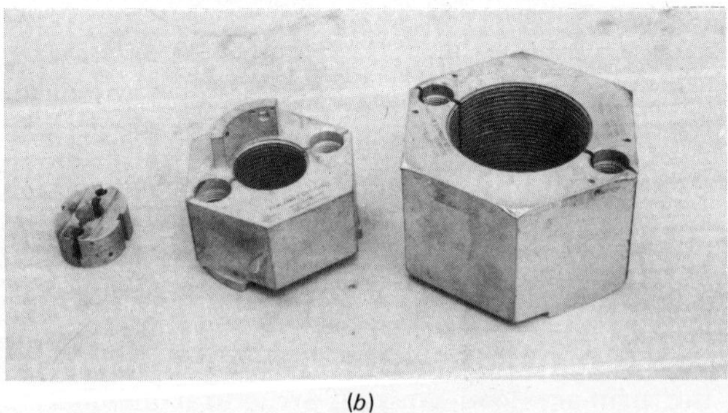

(*b*)

**Fig. 6-2.** Space shuttle frangible nuts. (*a*) Cutaway sections. (*b*) Photograph of actual frangible nuts.

at the forward attachment between it and the external tank. Due to the extreme skin temperature of 2,300°F (1,260°C) of the orbiter at this location during reentry from orbit, no operating doors were permitted in this area. This meant the forward separation system would have to provide a smooth outer moldline surface after separation from the tank. The same device would be used during the five separations of the orbiter from atop a modified Boeing 747 jumbo jetliner during the approach and landing test (ALT) program. Figure 6-3 shows the orbiter, moments after separation, during one of these drop tests to evaluate its atmospheric flying and landing characteristics. Because the envelope of separation trajectories for the orbiter from the 747 was broader than the envelope of the orbiter from the external tank, it was decided the separation planes of the orbiter's two aft attachments for the ALT program should be external of the orbiter's lower moldline to reduce the possibility of interference between the two vehicles during separation. This meant the 2.5-inch (63.5-millimeter) diameter frangible nuts (see Figure 6-2), normally used to attach and separate the orbiter from the external tank's two aft structural attachments, would not be utilized during the ALT program.

Figure 6-4 shows how three frangible bolts, like the one at the forward attachment, were used at each of the two aft attachments between the orbiter and the 747. Altogether, seven frangible bolts had to function simultaneously to structurally separate the orbiter. Note how the frangible bolts in the aft attachments are installed inverted in relation to the installation of the forward bolt. This arrangement enabled the separation plane to be located nearer

**FIG. 6-3.** Orbiter separating from Boeing 747.

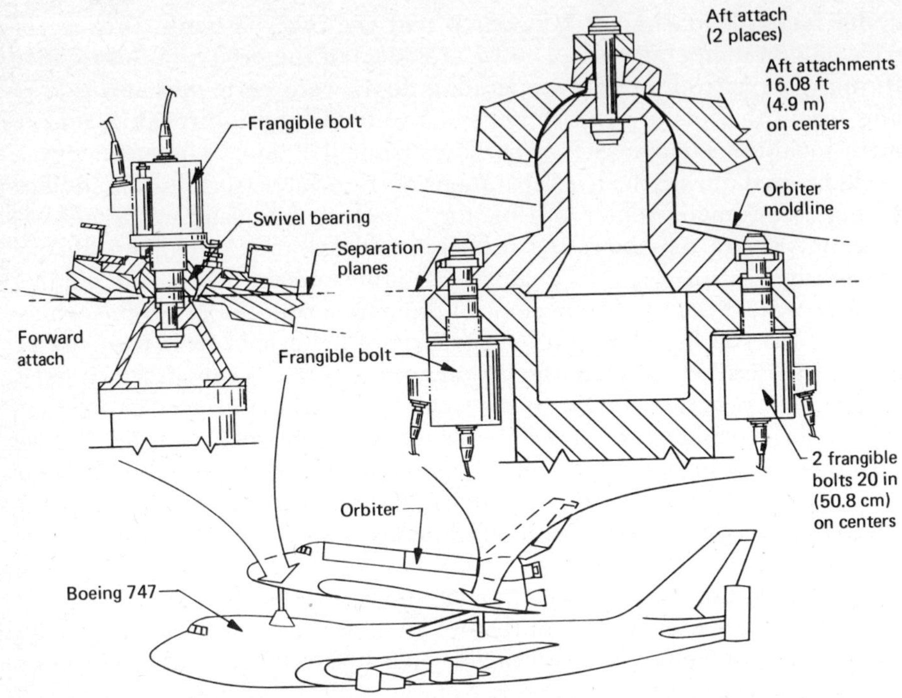


**FIG. 6-4.** Orbiter/747 separation interfaces.

the lower surface of the orbiter, otherwise the head of the two aft bolts would have interfered with the orbiter's lower surface, necessitating either dimpling the orbiter or lowering the separation plane. Inverting the frangible bolts alleviated both of these undesirable alternatives.

The internal configuration of the frangible bolt is illustrated in Figure 6-5*a*. It employs tandem pistons, each with a separate plenum chamber that is pressurized by redundant pyrotechnic pressure cartridges. These individual plenum chambers offer the added advantage of eliminating a potential single-point failure that is present when dual cartridges pressurize a common plenum above a single piston. If one cartridge should blow out of the bolt body, or leak, the pressure generated by the redundant cartridge could also vent through the same leak path of the faulty cartridge. The fracture or separation plane of the bolt is in the shank of the bolt at the base of the lower piston. The bolt shank is undercut externally at this location to provide a predetermined weak section that structurally fails in tension when either or both pyrotechnic cartridges are fired. The cartridge gas pressure in each of the plenums pushes down on the pistons and up on the bolt housing. These opposing forces are temporarily restrained by the strength of the bolt housing and shank. These forces increase as more cartridge propellant is burned, until finally the structural load-carrying capability of the weakened

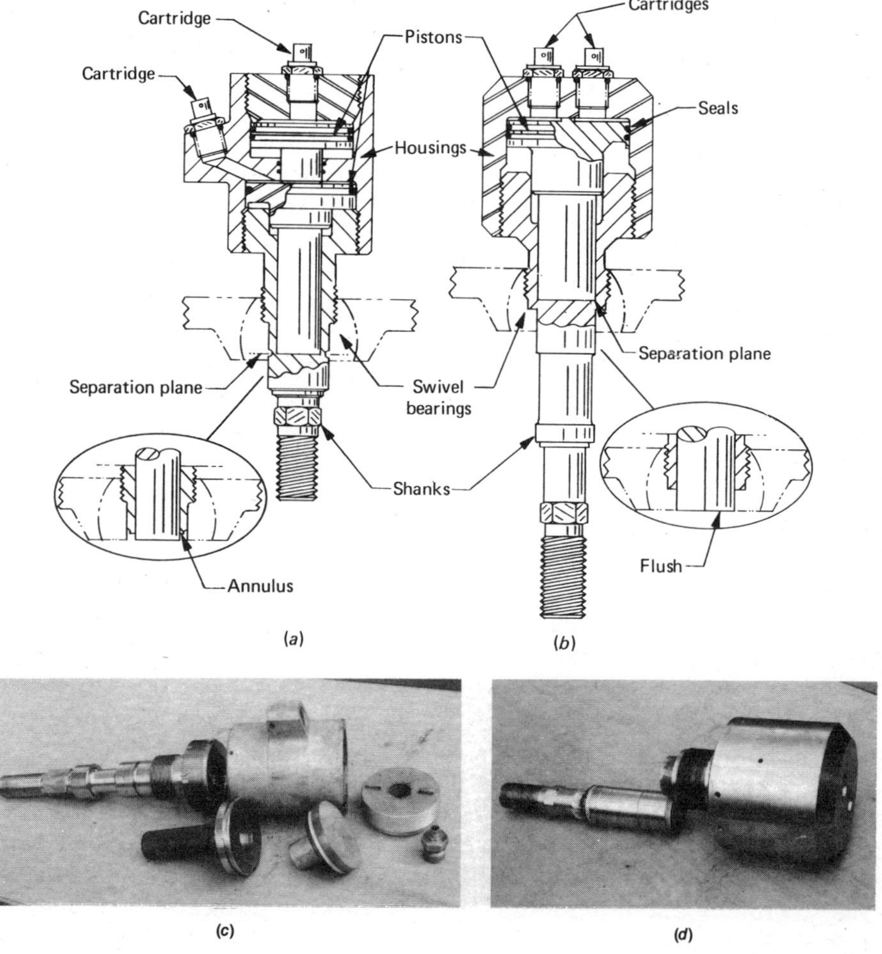


**FIG. 6-5.** Space shuttle orbiter frangible bolts. (*a*) Orbiter-to-747 frangible bolt. (*b*) Orbiter-to-external-tank frangible bolt. (*c*) Disassembled orbiter-to-747 frangible bolt. (*d*) Separated orbiter-to-external-tank frangible bolt.

shank is exceeded and it breaks. In the case of the forward attachment, the lower piston's extended travel partially plugs the hole created by the displaced lower shank, thus eliminating the need of a door to cover the bolt hole after separation.

Notice that there remains a small annulus on the outer surface of the orbiter that would be objectionable during reentry from orbit. Figure 6-5*b* illustrates a modification to the frangible bolt, for orbiter-to-external tank applications (see Figure 7-1), that changed the frangible mode of the shank from the previous tensile failure to shear failure. This design change eliminated the post separation annulus and met the revised orbiter surface smoothness criteria. Figure 6-5*c* pictures the disassembled tandem piston ALT tension bolt. Note the separation plane relief groove on the shank at left. Figure 6-5*d* pictures the external tank shear bolt after firing. Note the radial sheared surface on the shank on the end opposite the threads.

**Frangible Links** Frangible links are structural segments in series with one or more other structural members. One or more such assemblies may be used to support and/or retain a component that can be released at some later time. Frangible links may be designed to be broken by either pyrotechnic detonator or pressure cartridges.

From launch until earth orbit, the Apollo lunar module (LM) was supported at four locations inside a truncated conical adapter between the Saturn third-stage booster (S-IVB) and the command and service module (CSM) that housed the three lunar astronauts. In earth orbit the upper portion of the adapter was pyrotechnically severed at the base of the service module and at the LM attach plane. The cone was also simultaneously severed longitudinally into four panels that were immediately jettisoned, exposing the LM from the remaining stub adapter. The CSM then turned 180 degrees and docked with the LM. Astronauts installed electrical umbilicals between the CSM and the LM. After all the LM systems had been checked for proper working order, the astronauts would align the spacecraft on the proper lunar course and restart the big rocket engine at the base of the Saturn third-stage booster. This added velocity ("delta V" in space jargon) caused the spacecraft to escape earth orbit and proceed into the next phase of its journey, translunar injection. At the conclusion of this rocket boost, which lasted several minutes, the astronauts would structurally disconnect the LM from the stub adapter by simultaneously firing both detonators in each frangible link at the four LM/adapter attach points, as shown in Figure 6-6. At each location, two straps have one end attached to the stub adapter that straddles a LM fitting resting on a ball/socket interface with the adapter. The free ends of the straps are held together by a frangible link containing two hot-wire electric detonators (see Chapter 3). A LM tie-down bolt, between the frangible link and the LM fitting resting on the adapter, is preloaded to prevent the LM from breaking loose from the adapter during launch. When the detonators break each of the four frangible links, springs retract each disconnected strap away from the LM fitting. The tie-down bolt and block are attached to one of the straps and rotate away from the LM fitting with the strap. Reaction control system (RCS) jets on the service module are instantly fired to pull the CSM and LM away from the stub adapter. The Apollo spacecraft then continues on its long journey—coasting to the moon.

## Nonfrangible Devices

Nonfrangible separation/release devices generally require considerably less pyrotechnic material than comparable frangible devices performing the same function because the pyrotechnic material in the nonfrangible device provides

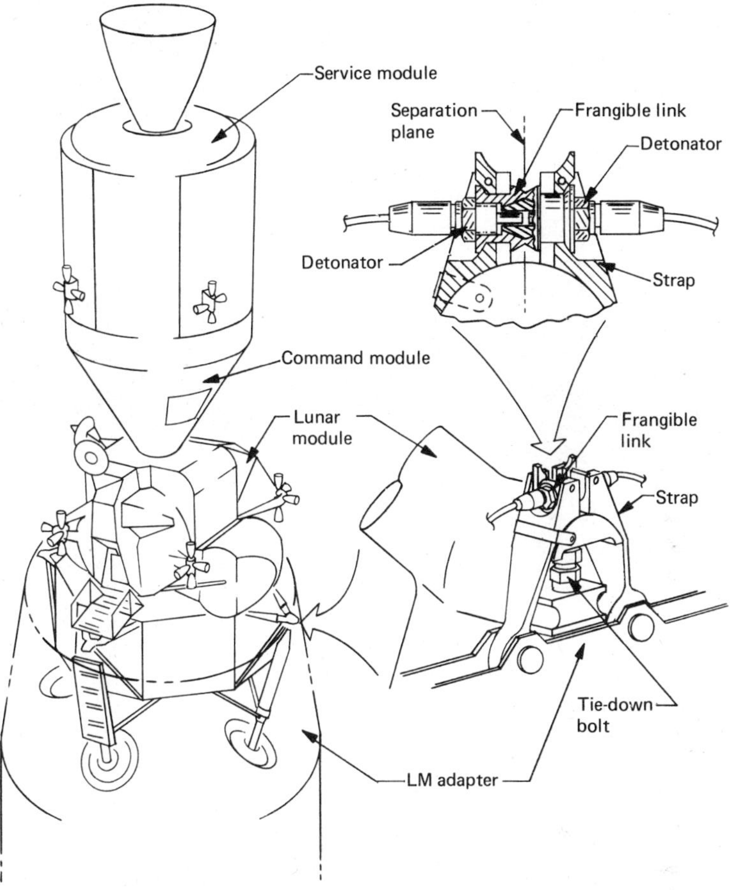


**FIG. 6-6.** Apollo lunar module frangible link.

the energy to "trip" a mechanical, load-carrying mechanism. The actual load being separated or released may be reduced by several stages (or toggles) before it gets to the pyrotechnic stage of the device. One big advantage of these devices is their relative reduction in pyrotechnic and mechanical shock while functioning compared to frangible devices. Their one disadvantage is their relative complexity. The mechanical stages (toggles) reduce their overall reliability, and they are generally more complex than their frangible counterparts.

**Release Nuts** These devices are available in many configurations that consist of three or four internally threaded segments held together by a collet to form the basic nut. A bolt or stud is threaded into the assembly to complete the classic nut/bolt fastener assembly. When the collet is made an integral part of a movable piston, as shown in Figure 6-7, the collet can be

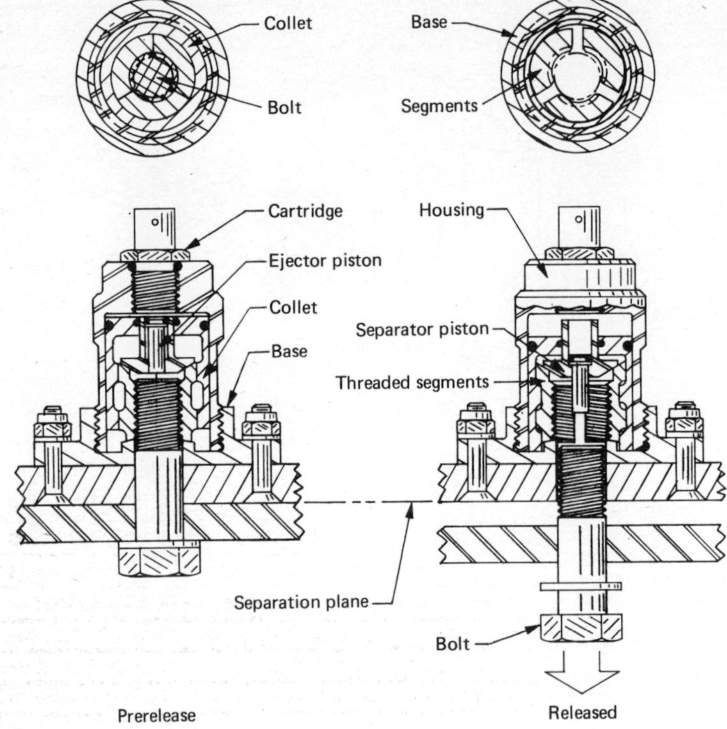


**FIG. 6-7.** Release nut.

moved axially to release the segments. If the bolt is significantly tightened (preloaded) or if there is a tensile load on the bolt when the collet piston is displaced, the tensile load at the threaded interfaces of the segments and the bolt will cause the segments to be radially displaced far enough to completely disengage the threads—thus releasing the bolt. The tangent of the thread form flank angle, i.e., 30 degrees for military standard, 7 degrees for buttress, etc., has a direct influence on the ratio of the axial load in the bolt to the radial forces in the segments—the greater the angle, the higher the radial load. The segment's radial load and the coefficient of sliding friction between the segments and the collet piston determines, to a great extent, the amount of pyrotechnic energy required to release the segments.

Other refinements include matching ramps between the base of the threaded segments and the base of the nut as well as matching ramps between the top of the threaded segments and the separator piston. These ramps help to displace the segments radially during actuation and thus ensure a more simultaneous disengagement of all the segments from the bolt or stud. They also prevent the segments from rebounding into the retracting bolt or stud. A shear (or lateral) load on the bolt at the time of release can impair its retraction. The addition of an ejector piston greatly reduces this possibility by ejecting the bolt from the release nut. All three pistons—collet release, separator, and ejector—are simultaneously exposed to the pyrotechnic cartridge pressure, but the function of each is automatically sequenced. This is accomplished by sizing the relative areas of each piston exposed to the cartridge pressure. For example, if the area of the ejector piston were larger than the collet piston, the radial load induced in the segments against the collet piston, by the ejector piston against the bolt, could stall the collet piston completely. The release nut depicted in Figure 6-7 is shown being fired by a single pyrotechnic pressure cartridge. Multiple cartridges can be incorporated into the release nut by simply canting the cartridge ports about the centerline of the release nut.

Properly designed, release nuts are highly reliable devices. They have been used to jettison the retro-rocket packages on the unmanned Surveyor lunar lander and the Viking Mariner lander, and are being used in numerous applications on the space shuttle.

## POWDER-ACTUATED FASTENING SYSTEMS

Not too many years ago, to fasten something to a concrete wall or steel beam would have meant a lot of time and expense in the preparation, e.g., placement of pneumatic hoses or electric extension cords to provide power sources for an air hammer or an electric drill. Considerable time was consumed preparing a hole for the installation of an expansion bolt in concrete or a bolt, washer, and nut in a steel beam or plate.

Shortly after World War II there was a sudden rise in construction costs, and a more efficient method of fastener installation was needed. Several manufacturers began to develop on an experiment that had its beginnings in England in the early 1900s. This concept was based upon the adaptation of the pistol firearm to propel a nail (instead of a bullet) into a base material (instead of through the air). From this effort there has evolved two principal concepts that dominate the powder-actuated fastening systems industry.

### Low-Velocity, Indirect-Acting Principle

The low-velocity, indirect-acting system consists of a tool weighing approximately 4 to 8 pounds (1.8 to 3.6 kilograms), a power load (cartridge), a piston, and a fastener as shown in Figure 6-8*a*. This classification of a powder-actuated tool as "low velocity" is based upon a procedure which is used to measure the velocity of a fastener as it is discharged from the tool as a free-moving projectile. Specifications require that the velocity measurements of the fastener be made 6.5 feet (2 meters) from the muzzle end of the tool and that the smallest commercially available fastener be used with the most powerful commercially available power load. A low-velocity tool by this definition is a tool which, when tested with a series of 10 tests in deliberate free flight, will produce an average velocity from the series not in excess of 300 feet (91.4 meters) per second. The velocity of a single test in

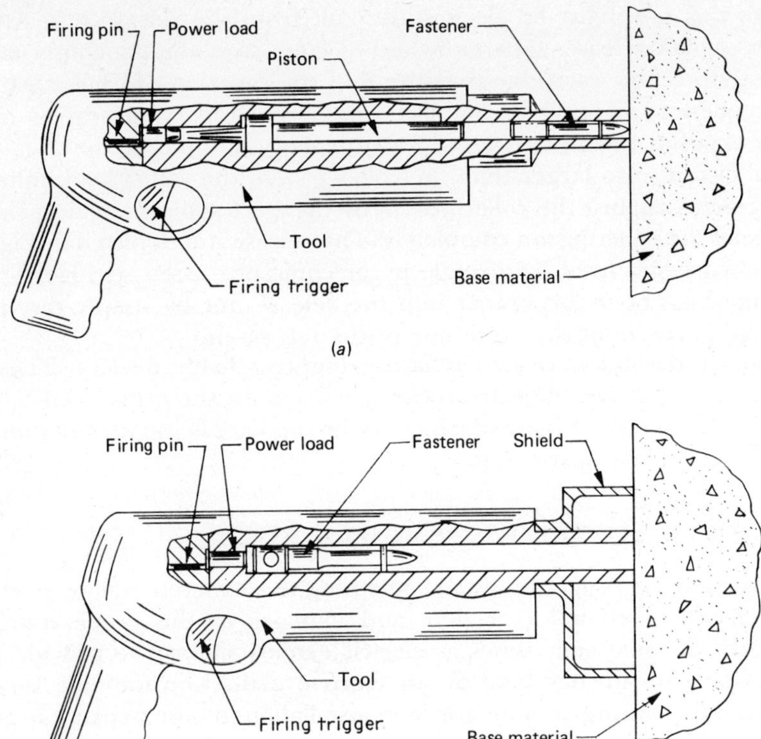


**FIG. 6-8.** Powder-actuated fastening system. (*a*) Low-velocity indirect acting principle. (*b*) Standard velocity direct acting principle.

that group cannot exceed 325 feet (99 meters) per second. Pressure developed by the confined burning of gunpowder provides the high force for driving the fasteners. Modern smokeless gunpowder burned in the open atmosphere burns comparatively slowly, but when confined burns very rapidly and produces high pressures. In some powder-actuated tools, these pressures can exceed 30,000 pounds per square inch (2,100 kilograms per square centimeter). The pressure developed by the confined burning of the power load gunpowder drives a piston, initially located near the power load, along the barrel of the tool until it strikes the fastener near the end of the barrel. The momentum from the piston when impacting the fastener is great enough to drive the fastener into the base material (concrete or steel).

Safety considerations require two distinct motions, in sequence, to fire the tool. First, the barrel end of the tool must be depressed with a force of several times the weight of the tool against a work surface at an inclination not exceeding 5 degrees from a right angle. Angles greater than 5 degrees could cause the fastener to ricochet from the base material, resulting in injury to the tool operator or to bystanders. While the tool is held against the work surface, the firing trigger must be squeezed. This safety feature closes a gap of approximately 0.25 inch (6.3 millimeters) between the power load and the firing pin. It is impossible to fire the tool unless the aforementioned actions are executed in sequence.

### Standard-Velocity, Direct-Acting Principle

The standard-velocity, direct-acting system consists of a tool weighing approximately 6 to 10 pounds (2.7 to 4.5 kilograms), a power load (cartridge), and a fastener, as shown in Figure 6-8*b*. The fastener is positioned next to the power load and is driven into the base material by the force of the power load acting directly on the fastener instead of a piston. A fluted (generally plastic) attachment on the pointed end of the fastener provides a frictional element to retain the fastener in place until the expanding gas drives the fastener through the barrel. It also aligns the shank of the fastener with the bore of the barrel, thus reducing the possibility of the fastener ricocheting from the base material.

The standard-velocity powder-actuated tool will produce an average velocity in excess of 300 feet (91.4 meters) per second as described in the test procedure of the low-velocity tools. Both the standard- and low-velocity tools have similar safety requirements and therefore they won't be repeated here.

The standard-velocity tool is shown with a protective shield or guard that protects the operator from heat scales from the fasteners and flying particles from the surface of the base material. These guards are usually made of 0.10-inch (2.5-millimeter) thick steel.

### Power Loads

Rim fire and center fire power loads are available in cartridge sizes ranging from 22 to 38 caliber (5.58 to 9.65 millimeters), a few of which are shown in Table 6-1*a*. End closure can either be crimped or wadded. Originally all power loads were wadded with cardboard or similar material. However, when the piston-type tools entered the powder-actuated fastening system, it was necessary to eliminate the wad residue because the combustion chambers of these tools are confined. The crimped cartridge case eliminated the need for the wad. As the piston-type tools increased in usage, power-load manufacturers produced standard-velocity loads in crimped cases. In some standard-velocity tools the crimped loads do not perform to some users' satisfaction as well as wadded loads.

In order to provide control over the penetration of a fastener, power loads are manufactured in incremental levels of power. Standards have been established for the power level and the identity of each power level for the loads used in powder-actuated tools. These standards were developed in the interest of safety to provide a uniform identification of the power load where

**Table 6-1.** Powder Cartridge Configurations and Load Identification

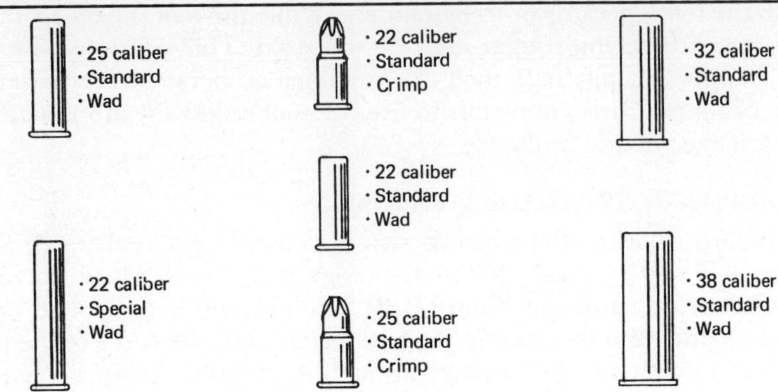


(*a*) Typical Powder Cartridge Configurations

| Cartridge | Color identification | | Nominal velocity* | |
|---|---|---|---|---|
| Power level | Case color | Load color | Meters per second (±13.5) | Feet per second (±45) |
| 1 | Brass | Gray | 91 | 300 |
| 2 | Brass | Brown | 119 | 390 |
| 3 | Brass | Green | 146 | 480 |
| 4 | Brass | Yellow | 174 | 570 |
| 5 | Brass | Red | 201 | 660 |
| 6 | Brass | Purple | 229 | 750 |
| 7 | Nickel | Gray | 256 | 840 |
| 8 | Nickel | Brown | 283 | 930 |
| 9 | Nickel | Green | 311 | 1,020 |
| 10 | Nickel | Yellow | 338 | 1,110 |
| 11 | Nickel | Red | 366 | 1,200 |
| 12 | Nickel | Purple | 393 | 1,290 |

*The nominal velocity applies to a 3/8 inch (9.53 millimeters) diameter, 350 grains (22.7 grams) ballistic slug fired in a test device and has no reference to actual fastener velocity developed in any specific tool.

(*b*) Powder Cartridge Load Identification

a competitor's load is used in a tool. Power levels for power loads are in increments numbered 1 through 12 as shown in Table 6-1*b*. The 12 levels are subdivided into two series: low series, 1 through 6, and high series, 7 through 12. The low series are identified with brass-colored cases while the high series have nickel-colored cases. One of six basic colors on the wad or head of the case further identifies the level of each power load in each series.

## Fasteners

The fasteners used in powder-actuated fastening systems are precision made from special steel, heat-treated for maximum strength, and plated for corrosion resistance. Ordinary nails, screws, and concrete nails will not work in powder-actuated tools. There is an almost endless variety of fasteners of numerous shank types, diameters and lengths, head diameters, head shapes, and thread sizes and thread lengths for almost any type of application, a small sample of which are illustrated in Figure 6-9. Headed fasteners are used for attaching wood sills, furring strips, and wood partitions. Threaded fasteners are used for attaching structural steel, sprinkling systems, and tracks. Eye pins are used for hanging false ceilings and as tie anchors. For rapid installation, special preassembled clips for lightweight drop ceilings and conduit installations are also available. Fasteners are manufactured from

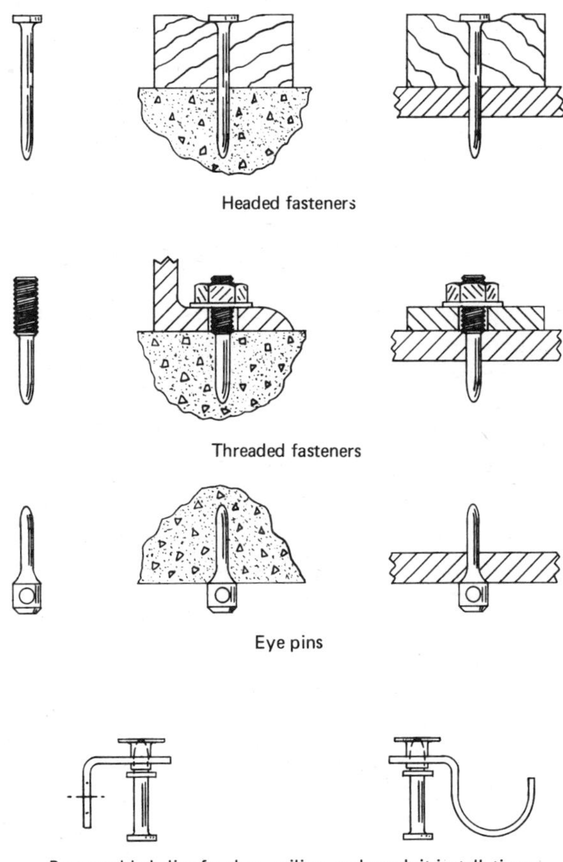


**FIG. 6-9.** Examples of powder-driven fasteners.

a special alloy wire that arrives from the steel supplier in coils. The wire is first drawn through a die to size and straighten and then cut to length and headed for headed fasteners or shanked for threaded fasteners. Threads are rolled to reduce local stress concentrations. Ballistic-shaped points are machined for easy penetration of the fastener into the base material. They are then heat-treated by the austempering process, which gives the fastener a combination of hardness and ductility to penetrate steel or concrete without plastic deformation or shattering. The resulting ultimate tensile strength of the fastener is up to 240,000 pounds per square inch (16,872 kilograms per square centimeter). A bright zinc electroplated finish followed by a chromate conversion process is applied to provide corrosion resistance and lubrication for easier installation. Finally, a guide(s) is installed on the shank for fastener/barrel alignment.

### Fastener Holding Power

The holding power of a fastener is determined by its shank diameter, which in turn is based upon the load on the fastener, the type of base material, and the driving capacity of the tool. Fasteners are made in shank diameters of 5/32, 11/64, 3/16, 7/32, and ¼ inch (3.97, 4.37, 4.76, 5.56, and 6.35 millimeters). In concrete applications, penetration equal to eight times the shank diameter will give optimum holding power. Structural steel applications require penetration where the full diameter of the shank exits out of the opposite side of the steel plate for optimum holding power. Fasteners with knurled shanks are made for use in steel where the fastener may be subjected to torque when nuts are applied. These knurled shanks will produce more holding power in steel under tension loads than will smooth shanks. They may break off sooner at lower shear and bending loads than will the smooth shanks because of the notch effect of the knurl.

**Fastener Holding Power in Concrete** A fastener is held securely in concrete by the compressive action of the concrete against its shank. As the shank forces its way into the concrete, it pushes forward and laterally, displacing its own volume. The displaced concrete is thus compacted around the shank. Both the concrete and shank are heated in the process, resulting in some fusion of small concrete particles with the shank surface. The compressive bond and fusion provides the holding power of the fastener. Sometimes a thin surface layer of concrete is broken (spalled) by the lateral forces of the entering shank. Spalling is, however, a normal result of the low shear strength of the concrete and does not affect the holding power of the fastener. Figure 6-10*a* diagrams the holding power of fasteners driven into concrete. The fastener's holding power is determined by the compressive strength of the concrete, depth of penetration of the shank, and diameter of the shank. Shank diameter is selected according to the holding power required from the fastener and the compressive strength of the concrete

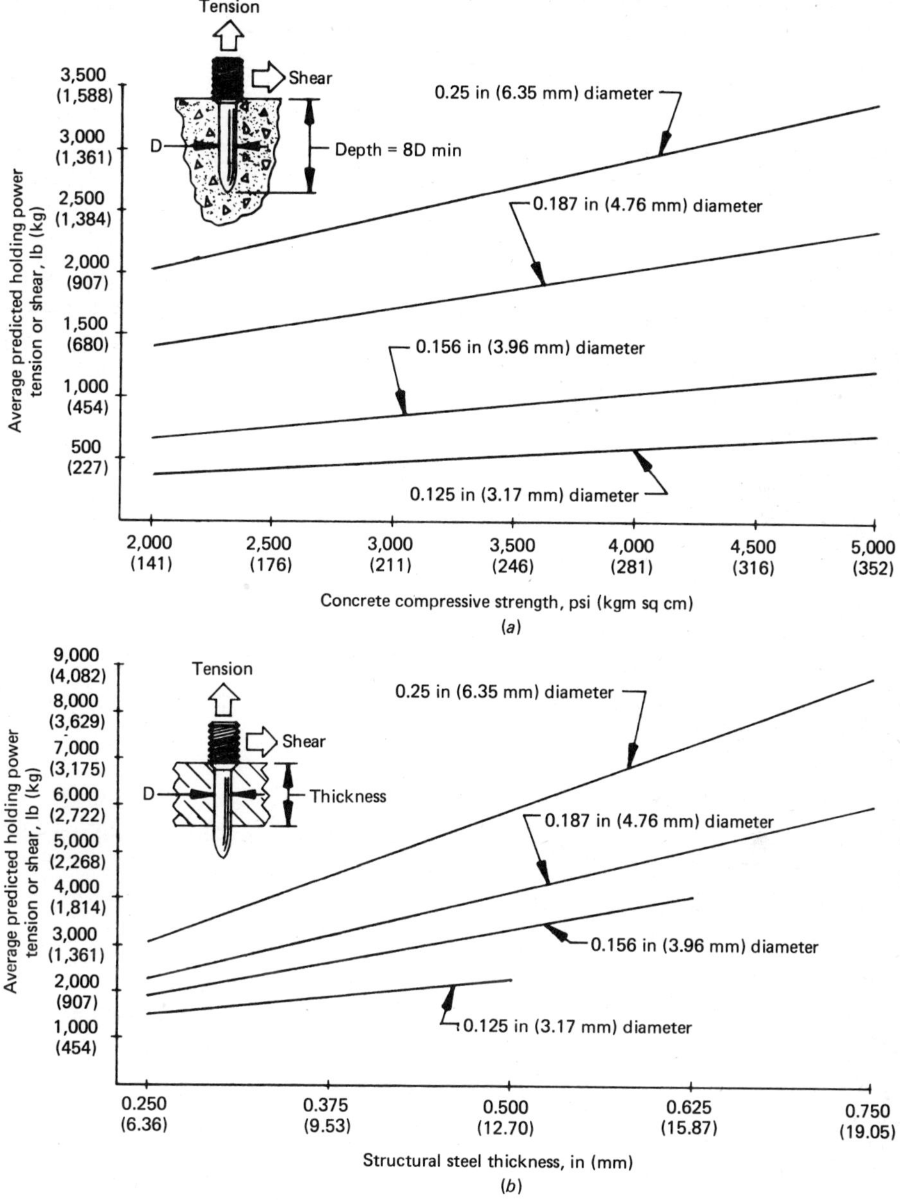


**FIG. 6-10.** Holding power of fasteners driven into concrete and steel. (*a*) Concrete. (*b*) Steel.

into which it is being driven. The larger the diameter, the greater the holding power. In the diagram, all fasteners were driven into the concrete to a depth equal to eight shank diameters.

**Fastener Holding Power in Steel** The holding power of a fastener driven into steel is due to the compressive force of the steel-base material acting on the shank of the fastener. As the shank is forced into the steel, it pushes the steel aside, simultaneously compressing and work hardening it. The compression and friction heat the interfacing surfaces of the steel and fastener shank. The compressed steel pushes the heat softened surfaces together to form a tight bond. The steel cools rapidly after the fastener has seated and produces a tightly squeezing ring of hard steel around the shank of the fastener. Tests show the steel base material is strengthened near the fastener shank due to hardening. Figure 6-10*b* diagrams the holding strength of fasteners driven into steel plates. The data presented are for smooth shank fasteners. The conventional configuration of fasteners designed specifically for use in steel have knurled shanks. The limitations of knurled shanks were discussed earlier.

### Limitations of Base Material

Unsuitable fastener base material can be classified into three categories:

**1.** TOO HARD Fastener will not penetrate and could possibly ricochet or break.
*Examples:* Hardened steel, welds, cast steel, marble, spring steel, and natural rock.

**2.** TOO BRITTLE Material will crack or shatter and fastener could ricochet or pass completely through base material.
*Examples:* Glass, glazed tile, brick, or slate.

**3.** TOO SOFT Material does not have the characteristics to provide holding power and fastener could pass completely through.
*Examples:* Wood, plaster, drywall, composition board, and plywood.

## CARTRIDGE STARTING SYSTEMS FOR DIESEL ENGINES

One of the primary requirements for diesel engine starting is fast, reliable cranking. The problem of cranking, which is most acute at sub-zero temperatures, is not due to deficiencies in the cranking motors but more in the difficulty of storing the energy for the starter. The three primary energy sources are electrical, compressed air, and hot gas that are stored in batteries, tanks, and pyrotechnic propellant cartridges respectively. Figure 6-11 shows the low-temperature performance capabilities of these three power sources. At −15°F (−26°C), batteries have less than 50 percent of their room-

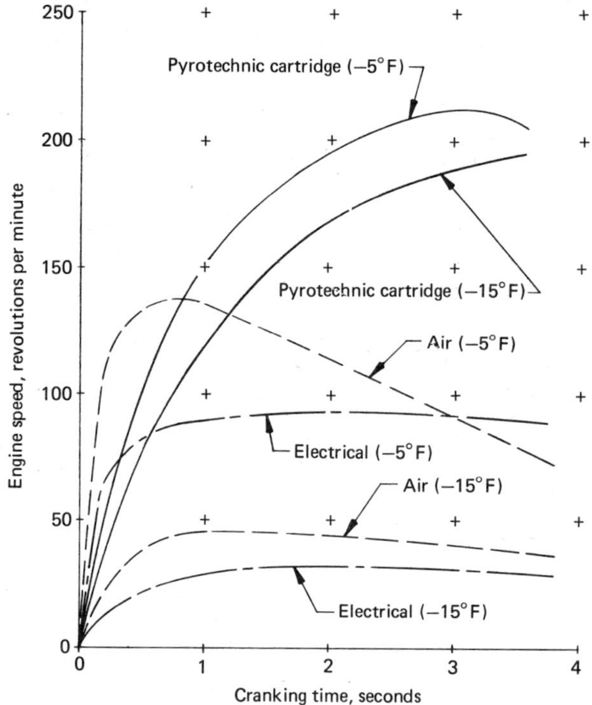


**FIG. 6-11.** Low-temperature performance capabilities of diesel starting systems.

temperature power. Compressed air is difficult to store and cold tanks and tubing reduce the available pressure at a time when higher tank pressure is normally required to compensate for the lower temperature. Gas pressure generated by a pyrotechnic cartridge is much less affected by the lower temperature because it provides its own compensating heat.

## The Starting System

The storage components required for the two most efficient low-temperature energy sources, compressed air and solid pyrotechnic propellant, can be combined into a compatible dual power source diesel engine starting system. The system is intended primarily to start diesel engines with displacements in the 400- to 900-cubic-inch (6.56- to 14.75-liter) range. Even though cartridge-starting diesel engines are relatively new, the utilization of low-flame-temperature solid-propellant cartridges for turbine starting applications is well established. Since World War II, millions of pyrotechnic cartridges have been used for starting military jet aircraft engines. It is this basic pyrotechnology that is being converted to commercial diesel engine starting applications.

The starting system employs gas from either of the two separate power sources as shown in Figure 6-12. The primary power source, compressed air, is stored in a tank at 150 pounds per square inch (10.5 kilograms per square centimeter) pressure. The secondary power source, hot gas, is generated by a pyrotechnic cartridge. The operator may activate either power source from separate controls located in the vehicle cab or on the engine control panel. A solenoid control valve is located near the tank along with a check valve that prevents hot gasses from the pyrotechnic cartridge from damaging the solenoid valve. After the compressed air has passed through the solenoid valve and the check valve, it is directed, via a manifold, into a vane-type starting motor mounted adjacent to the diesel engine's flywheel. Vane-type starting motors work well with low-pressure air, and routine engine starts can be made with compressed air stored at the pressure level that is compatible with vehicle air brakes and shop air. The depleted tank pressure is normally replenished by the diesel engine's compressor shortly after it is started.

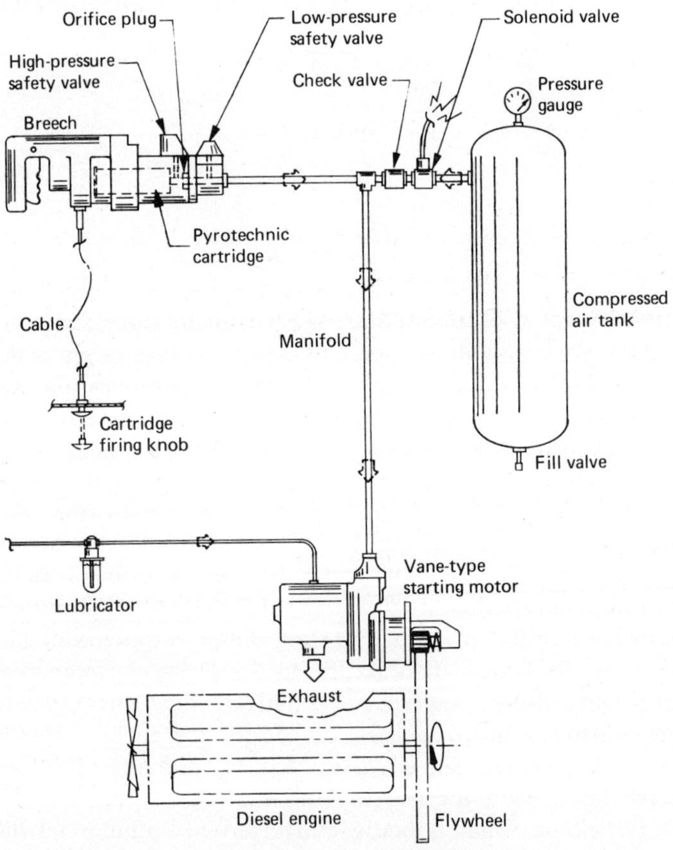


**FIG. 6-12.** Schematic of diesel engine dual starting system.

A cartridge start begins by placing a pyrotechnic cartridge in a breech that has a firing pin release cable attached to a firing knob in the vehicle cab or on the engine control panel. When the firing knob is pulled, a spring-loaded firing pin in the breech is released. The pin strikes a percussion primer (see Chapter 3) pressed into one end of a cartridge case. The flash from the primer ignites a priming mix (see Chapter 2) that in turn ignites the main propellant charge. Even though the engine usually starts within the first several turns, the entire volume of propellant gas produced must flow through and be exhausted from the vane motor. These exhaust gases, even though considered clean burning, contain small amounts of unburned solids and ash. These solids can collect on the interior of the vane motor and cause the rotor vanes to stick. To prevent sticking and clogging, provisions are made to inject a small amount of lubricant directly into the vane motor. This lubricant provides a flushing medium each time a cartridge is fired and ensures a smooth-running vane motor even after many hundreds of cartridge starts. Because the vane motor turns at several thousands of revolutions per minute, its output shaft drives a reduction gear train before engaging the geared flywheel of the diesel engine.

### System Components

The following discussions describe the dominant features of the starting system's major components and parametrically compares the performance characteristics of the two energy sources.

**The Tank** The compressed-air tank generally has a capacity of approximately 33 gallons (125 liters) with a burst pressure of 4 times the rated capacity of 150 pounds per square inch (10.5 kilograms per square centimeter) and is proof pressure-tested at 2.25 times its rated operating pressure. A pressure gauge is installed to indicate if an adequate quantity of air is stored for a proper engine start. Quite often cartridge starts are necessary because a leak in the tank has lowered the pressure to where adequate vane-motor speed cannot be achieved to start the diesel engine. An auxiliary fill valve provides a means of employing shop air to raise the tank pressure sufficiently to start the engine.

**The Breech** The breech acts as the cartridge combustion chamber and is capable of operating at pressures up to 3,000 pounds per square inch (211 kilograms per square centimeter), while exposed to temperatures up to 2,000°F (1,093°C), and can tolerate the corrosive effects of propellant gases and other products of combustion. The rear section contains a firing mechanism that can be remotely fired from the cab or engine control panel. Breeches now in use separate near the rear end and are joined by a "pressure-cooker" type of flanged joint that locks together when the two pieces are joined and rotated approximately 45 degrees relative to one another. A firing pin spring within the firing mechanism is automatically cocked during this locking sequence. To prevent cartridge firing during the chambering process, the breech incorporates a lockout mechanism that prohibits the firing pin from being released when the breech halves are not completely closed and locked. A close-tolerance orifice plug is installed near the exit end to maintain proper breech pressure as the cartridge propellant deflagrates.

**The Pyrotechnic Cartridge** Figure 6-13*a* shows a typical starter cartridge consisting of a standard double-seamed tin can which serves as a cartridge case. A percussion primer is pressed into the center of the ignition end. The fire from the primer ignites a small quantity, approximately 10 grams, of priming mix that is sufficient to ignite the entire propellant charge instantly. The opposite end of the can is coined so it will rupture under low pressure and allow emission of gases when the propellant charge (Figure 6-13*b*) is ignited. The shape of the propellant charge is so configured that as the burning surfaces recede, the burning area remains relatively constant. That is, as the burning area on the outer surface decreases, the burning area on the inside surface increases at approximately the same rate. Because of a

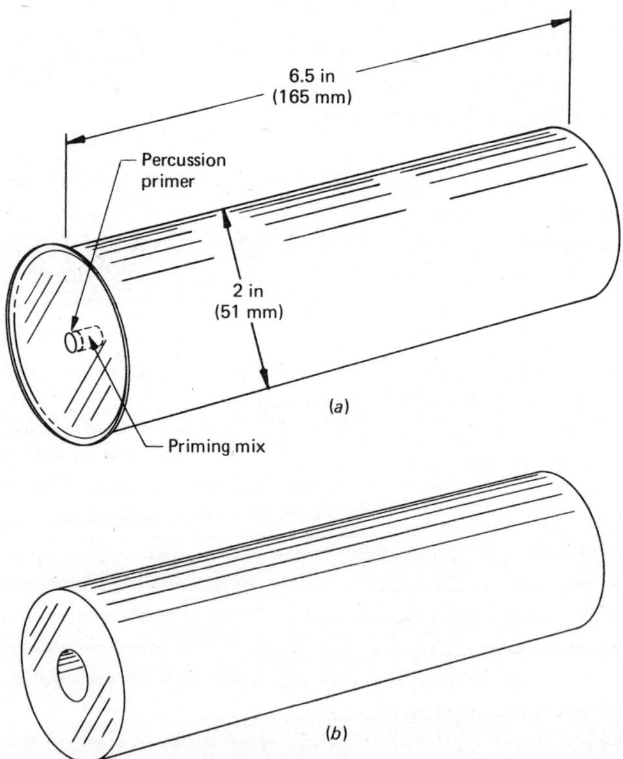


**FIG. 6-13.** Diesel engine pyrotechnic starting cartridge. (*a*) Cartridge. (*b*) Propellant charge.

rather large length/diameter ratio, the effects of the burning cylindrical ends are negligible.

The solid propellant used in diesel engine starting cartridges is classified as low energy, having a linear decomposition rate of less than 0.01 inch (0.25 millimeter) per second and a flame temperature of about 2,000°F (1,093°C). The most common solid propellants used in military rocket applications have decomposition rates 5 to 20 times faster and flame temperatures 2 to 3 times greater. The propellant charge is composed of ammonium nitrate which is the oxidizer, and a rubber compound which serves as both the fuel and binding agent. The outstanding characteristic of an ammonium nitrate propellant, beside its low combustion temperature, is a yield of relatively clean gases consisting of nitrogen, hydrogen, carbon dioxide, water vapor, and small amounts of carbon monoxide and methane. During manufacture, the materials are mixed into a homogeneous powder, then either pressure-molded or extruded to a final density similar to a common rubber compound. After compacting, the propellant charge may be machined or sawed to its final configuration. It can be exposed without damage to temperatures from −60°F to 180°F (−51°C to 82°C). Storage data include periods in excess of 5 years in warehouses without change in propellant burning characteristics.

There are two basic methods in which propellant gases may be used in diesel engine starting applications: (1) to use the hot gases as they are being generated and (2) to duct them into a high-pressure storage tank, permit them to cool, then use them whenever an engine start is required.

The initial approach was chosen since it takes advantage of the available thermal energy. The latter approach would essentially be an alternate way of charging a compressed-air tank. It does, however, have the advantage of permitting the use of a conventional cold-air starting system in conjunction with a high-volume filter to ensure clean gases. The following comparative volume discussion exposes the main objections to the tank charging method of utilizing propellant-generated gases. The propellant charge illustrated in Figure 6-13*b* weighs 0.87 pound (0.39 kilogram) and has a volume of 17 cubic inches (278 cubic centimeters), and when burned would charge a 55-gallon (208-liter) tank to 150 pounds per square inch (10.5 kilograms per square centimeter) at a temperature of 2,000°F (1,093°C). If the same volume were permitted to cool to room temperature, the gas would only charge a 9-gallon (34-liter) tank to the same pressure. At room temperature, about 25 percent of the original propellant would have condensed into a liquid. The remaining reduced volume of compressed gas would yield only a little over 1 second of normal cranking speed. Figure 6-14 plots the two cases above as well as all volumes and temperatures in between. The aforementioned 55-gallon tank charged with 2,000°F gas at 150 pounds per square inch pressure represents the maximum theoretical volume obtainable from the propellant charge illustrated. An actual cartridge tank-charging system with an efficiency this high would require internal insulation to reduce heat losses and exotic materials to withstand the high temperatures. A tank of 33 gallons (125

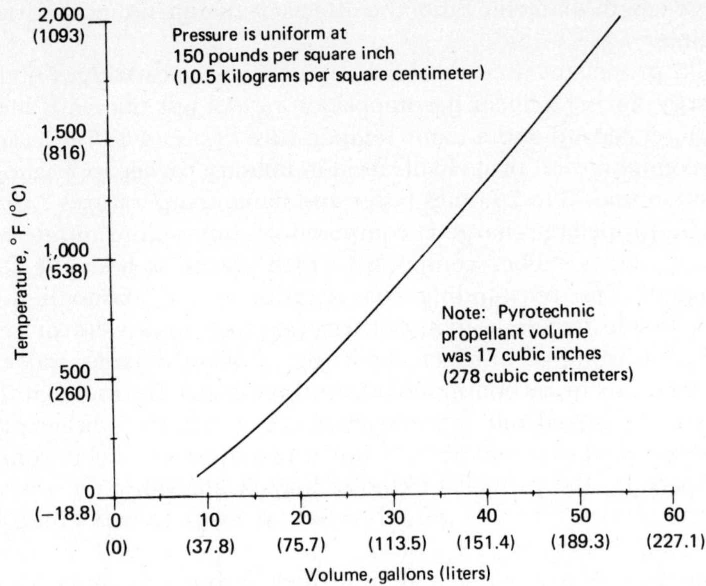


**FIG. 6-14.** Pyrotechnic propellant gas temperature versus volume with uniform pressure.

liters) capacity and a gas temperature of 1,000°F (538°C) with the same 150 pounds per square inch (10.5 kilograms per square centimeter) pressure would be a practical compromise and would best represent the equivalent input conditions of current air-starting systems. It should be remembered, though, that the propellant gas is not stored in a tank, but is used as it is generated to take full advantage of its thermal energy. Figure 6-14 shows that at an average system gas temperature of 1,000°F, the available gas volume is approximately 3.5 times what it would be if the gas were cooled to room temperature.

Actual gas pressure at the vane motor is a function of gas temperature, the reacting load, and the rate at which the gas enters the vane motor. Two of these parameters, gas input temperature and flow rate, can be varied within certain limits. The rate at which the cartridge propellant charge produces the required gas is a function of the operating pressure within the breech, the geometry of the propellant charge, the basic propellant formulation, and to a lesser extent, the temperature of the pyrotechnic propellant charge at the time it is ignited.

When ignited, propellant charges burn only on their exposed surfaces. The burning surfaces recede normal to themselves and at a constant rate if the burning surface area remains constant. The burning rate of any particular propellant formulation is a direct function of the pressure within the combustion chamber (breech). This pressure is controlled by a fixed-diameter orifice plug installed in the output end of the breech. For example, if the orifice were sized for 500 pounds per square inch (35 kilograms per square centimeter) operating pressure, the propellant would burn at a rate of 0.06 inch (1.5 millimeters) per second and would produce gas at a rate of 0.15 pound (0.068 kilogram) per second. If the orifice were sized to operate at 1,500 pounds per square inch (105 kilograms per square centimeter), the rate of burning would be 0.10 inch (2.5 millimeters) per second and produce gas at a rate of 0.25 pound (0.113 kilogram) per second. The later illustration approximates the operating conditions of present-day diesel engine cartridge starting systems. The propellant charge starts burning within 50 milliseconds after primer ignition and continues to burn for approximately 3 to 4 seconds. Figure 6-15 plots both breech and vane motor inlet pressures as a function of the total propellant charge burning time.

## Comparative Low-Temperature Cranking Performance

At subzero temperatures, when engine resistance to cranking is greatest, a cartridge starter exhibits the following four key advantages over the compressed-air starter.

**High Starting Torque** At normal temperature, the starter inlet pressure required to crank an engine is relatively low, i.e., 50 to 70 pounds per square inch (3.5 to 5 kilograms per square centimeter). At low temperatures the required pressure may go as high as 225 pounds per square inch (15.8 kilograms per square centimeter), a manifold pressure that can be easily met by a gas-generating pyrotechnic cartridge. Maximum manifold pressure is limited by the setting of the low-pressure safety valve usually located in the output end of the breech. By contrast, the maximum starting torque for a compressed-air starter is limited by the tank air pressure, usually 150 pounds per square inch (10.5 kilograms per square centimeter). Figure 6-16 is a plot of torque versus starter revolutions per minute for these two comparative systems.

**High Cranking Speed** It is generally accepted that high cranking speed increases starting reliability. Engine compression and, therefore, heat of

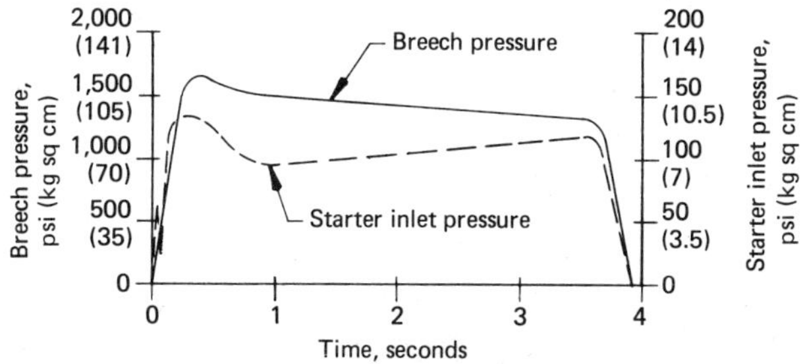


**FIG. 6-15.** Simultaneous breech and starter inlet pressure/time profiles.

compression is higher at fast crank speeds because there is less time for leakage past the piston and valves. There is also less time for heat to transfer to metal engine components. Cartridge starters are sized for fast cranking at low temperatures. Comparative horsepower versus starter speed is also presented in Figure 6-16.

**Constant Cranking Speed** During their gas-generating period, cartridges crank at relatively constant speed. This is in contrast to compressed air since pressure decays rapidly in a fixed-volume storage tank, and cranking power decreases proportionally. During low-temperature starting, the first few engine revolutions are utilized for fuel priming and drawing ether starting fluid into the engine cylinders.

**Line Freezing** During cold weather the inability to open the solenoid control valve due to water vapor freezing in the line is fairly common in compressed-air starting systems. This is especially true when shop air is used to recharge air tanks and the compressor inlet is located in a humid area. Freezing is not a problem with cartridge starting since manifold input gas temperatures are approximately 1,000°F (538°C) and the initiation system is independent of air.

## Cartridge Starting Limitations

In spite of the many advantages of cartridge starting over compressed air, there are two primary reasons why the cartridge crank cycle should be held to a minimum. The first is economic—as cartridges are "one-shot" expendable components, their replacement costs are considerably higher than recharging a compressed-air tank. The second is that all cartridge gases must flow through the vane starter motor even though the engine may start in many cases within the first few revolutions. If an overrunning starter drive is used, the drive remains engaged with the engine flywheel throughout the

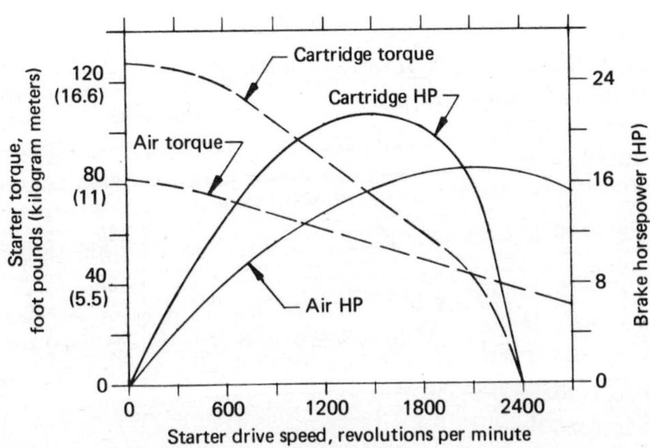


**FIG. 6-16.** 150-psi air and pyrotechnic cartridge torque and horsepower performance characteristics.

cranking cycle. If a friction clutch-type starter drive is used, the drive disengages from the engine flywheel. In both of these cases, when the engine starts the cranking load is removed from the starter and the starter runs to free speed and remains there until the cartridge gas is expended. Excessive duration at free speed increases the wear rate on the rotor vanes. Conversely, with the compressed-air starting system, the solenoid-controlled air valve is opened only until the engine starts, thus minimizing the wear on the starter rotor vanes.

## THRUSTERS, RETRACTORS, AND PIN-PULLERS

There is a family of pyrotechnic devices in which each consists of similar basic components, i.e., a piston within a housing, a shaft attached to the piston that also extends through the housing to the exterior, and a pyrotechnic, gas-generating cartridge threaded into the housing.

If the gas generated by the cartridge is ported to the gross side of the piston (opposite the shaft), the gas pressure will cause the shaft to extend further outside the housing. Such a device is called a *thruster*.

Conversely, if the gas generated by the cartridge is ported to the net side of the piston (on the same side as the shaft), the gas pressure will cause the piston to move in the opposite direction, so that the extended shaft is retracted into the housing. Depending on the configuration of the extended end of the shaft, such a device can be either a *retractor* or a *pin-puller*. A better understanding of all three types of piston devices can be had by a detailed description of actual units from each of the three groups.

### Thrusters

During the space shuttle orbiter's approach and landing test (ALT) and orbital flight test (OFT) programs, the two astronauts sit in ejection seats that provide them escape capability in the event of an emergency. Prior to initiating the seat ejection rocket-catapults attached to their ejection seats, two dual-panel overhead hatches had to be severed (see Chapter 4) and jettisoned from the orbiter to provide an opening through which the astronauts could eject. Since the severance systems did not impart enough momentum to the hatches to reliably jettison them, an auxiliary device was required to perform this function. The crew escape pyrotechnic schematic (see Figure 7-4) identifies these two devices as *hatch jettison thrusters* (at the lower center of the schematic).

Figure 6-17 illustrates the thruster, and as can be seen, it incorporates a few added features besides those of the aforementioned basic thruster components. The rod end at the left end of the piston shaft attaches to the jettisoned hatch while the right spherical bearing attaches the thruster to the ejection seat rail support structures. Figure 7-5*a* and *b* shows the two spent thrusters nestled between the ejection seat rails. When the thrusters are installed between the hatches and the rail support structures, the crew

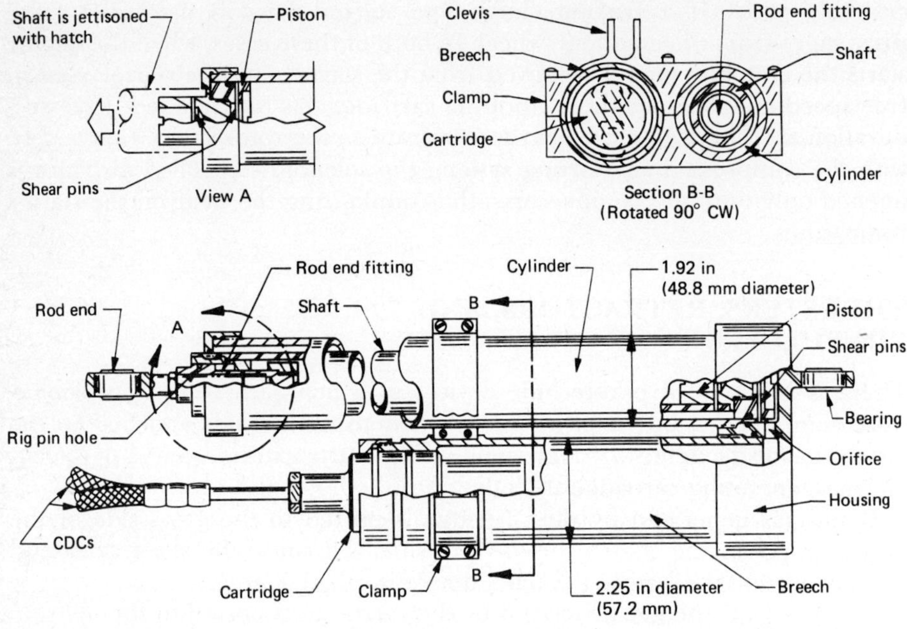


**FIG. 6-17.** Space shuttle orbiter's crew escape hatch jettison thruster.

module expands as the orbiter ascends into orbit. This expansion, of approximately 0.25 inch (6.2 millimeters), must be accommodated by the thrusters or else undue stresses will be imparted to the hatches, thrusters, and ejection seat rail support structures. During thruster installation, a rig pin is inserted through the rig pin hole, to temporarily lock the rod end fitting midway in its free-play travel. With the housing bearing attached to the ejection seat rail support structure, the thruster's nominal length of 23 inches (58.4 centimeters) can be adjusted by turning the rod end in or out of the rod end fitting to shorten or lengthen the shaft according to each installation requirement. When the installation is complete, the rig pin is removed and the expanding and contracting crew module will not impart loads into the thruster because the rod end fitting is capable of "breathing" as the crew module expands and contracts during the orbiter's mission into orbit and return. At the opposite end of the shaft is the piston with a shear pin anchoring it to the housing. This shear pin prevents the piston from moving with the expanding crew module instead of the rod end fitting that could cause excessive wear on the piston O ring. The shear pin is sheared during the initial stroking of the piston when the pyrotechnic cartridge is fired.

The thruster is capable of exerting a maximum jettison force of 12,000 pounds (5,443 kilograms) along a stroke of 18.75 inches (47.63 centimeters). At the end of the thruster's stroke the piston stops and the piston shaft is separated from the piston by shearing a pin that fastened the two components together. Pressure from the pyrotechnic cartridge, over the piston area, imparts a force on the piston that is transmitted to the shaft, and hence the hatch, by an interfacing shoulder around the piston annulus that contacts the shaft. The momentum the rotating hatch imparts on the piston shaft far exceeds the resisting shear capability of the pin. The piston could have been made the same diameter as the shaft, enabling it to exit at the end of the cylinder along with the shaft. This design option was prohibited by the requirement that all products of combustion and debris had to be contained within the thruster. The retention of the piston and the addition of the shear pin fulfilled this requirement. Figure 7-3*c* shows the orbiter's two jettisoned hatches at the left side of the picture. A close look at the upper hatch reveals the thruster's piston shaft extending upward from it.

The gas-generating pyrotechnic cartridge is initiated by redundant CDC lines (see Chapter 4) that make up the crew escape system's energy transfer system (ETS) (see Chapter 7). Generated gas from the pyrotechnic cartridge pressurizes a relatively large plenum chamber in the breech. At the same time, the pressure is also transmitted to the gross side of the piston through an orifice in the common bulkhead in the housing between the breech and the thruster cylinder. The effect of the large plenum is to reduce the initial high forces into the hatch, along with high accompanying reaction loads into the ejection seat rail support structure. The thruster's relatively light weight of 8.2 pounds (3.7 kilograms), including the pyrotechnic cartridge, is quite remarkable for a thruster of this size.

The split clamp serves two main functions: providing stability to the cantilevered end of the breech and forming an integral clevis to which attaches an antirebound device that prohibits the thruster from rebounding back into the ejection seat trajectory envelope after the hatch has been jettisoned with the thruster's shaft attached.

These are but a few of the many characteristics of pyrotechnic thrusters. Other features include pins extending from the piston through the orifice into the cartridge plenum that meters the rate of gas flow as a function of piston stroke. This pyro-technique can control both the piston load (force) and rate of extension in an infinite number of variations by simply varying the pin diameter along its length. For higher functional reliability, some thrusters have redundant cartridges. When only one cartridge fires, the thruster's performance is considerably less than when both cartridges fire. One method of approaching uniform thruster performance, under this extreme variation in pyrotechnic energy, is to build an inner piston within the main piston attached to the shaft. The volume behind the inner piston is filled with crushable honeycomb of either thin metal foil or paper. The purpose of the inner piston is to exert an overwhelming force on the restraining honeycomb when both cartridges fire, causing the honeycomb to crush, and thereby increasing the initial volume of the plenum resulting in a lowering of pressure and force on the piston. Conversely, when only one cartridge fires, the amount of crushing force on the honeycomb is greatly reduced, thereby maintaining a much smaller plenum volume. An experienced pyrotechnic engineer can nearly duplicate the performance curves of single and dual cartridge firings after a few calculations and several test firings to adjust the crushing characteristics (density) of the honeycomb. Computer programs have been developed that refine the analytical phase of thruster development even further.

## Retractors

Once the orbiter arrives in orbit and all systems have been checked, its long-awaited purpose in being really just begins—to deploy and retrieve payloads in earth orbit. The system that provides the orbiter with this capability is called the "remote manipulator system" (RMS). It consists of a manipulator arm (boom) approximately 53 feet (16 meters) long, with one end attached to the forward end of the orbiter's payload bay longeron, as shown in Figure 6-18. The manipulator consists of four segments, connected by actuator-driven pin joints with a collet device built into the free end that has the capability of engaging a variety of tools and payload handling end effectors that can be exchanged in orbit. The RMS is controlled from the crew

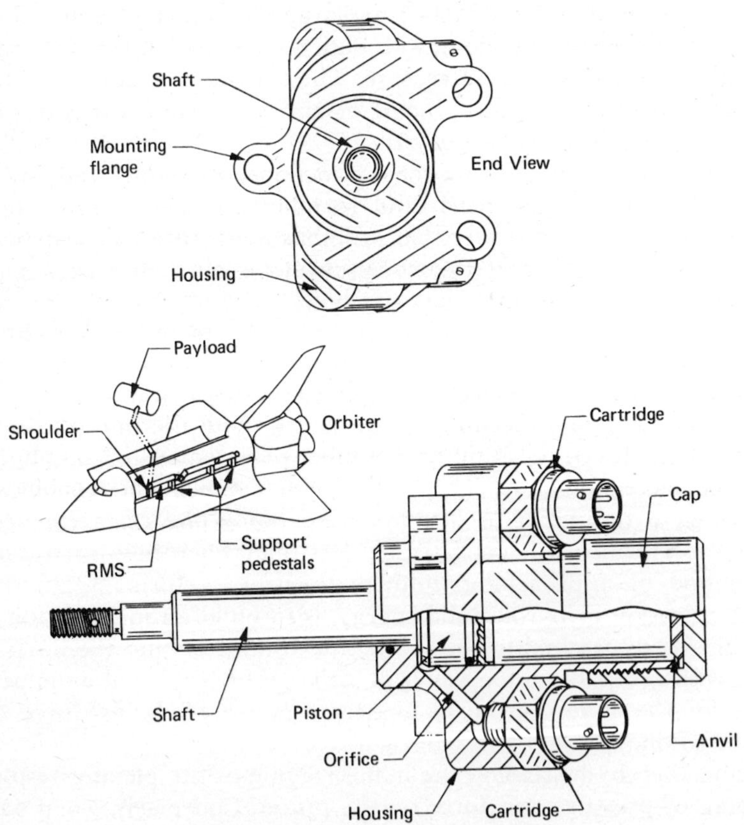


**FIG. 6-18.** Space shuttle orbiter's RMS jettison retractor.

compartment by an astronaut who has undergone specialized training to earn the title of payload specialist. With the aid of direct viewing windows and closed-circuit television cameras, located within the payload bay and on the manipulator, the payload specialist actuates switches and control handles on a control panel to maneuver the manipulator to any desired position required to deploy or retrieve a payload from within the orbiter's payload bay. Some payloads require the use of a second RMS on the opposite side of the payload bay. In operation, the dual manipulators resemble somewhat the articulating arms of a praying mantis insect.

During launch, the RMS is supported along the longerons by three support pedestals. They and the RMS shoulder lean inward and are sandwiched between the payload and the payload bay doors. Once in orbit, the doors are opened and the RMS shoulder and pedestals are rotated outboard approximately 31.5 degrees to provide adequate clearance for payload removal. When the RMS shoulder and pedestals have been rotated outboard, they extend outside the envelope of the closed payload bay doors. If the RMS shoulder or any of the three pedestals should fail to rotate back within the envelope on command, the orbiter could not survive the subsequent reentry environment without the payload bay doors closed.

Built within the base of the manipulator's shoulder and each of the three pedestals are separation systems consisting of mechanical toggle mechanisms that are tripped by four pyrotechnic retractors. There are also four pyrotechnic cartridge-actuated, electrical wire bundle umbilical guillotines at the RMS separation interfaces. Pyrotechnics were chosen because they were the only type of device that could function fast enough to release all four mechanisms and sever all four umbilicals simultaneously. In the case of a jammed deployed manipulator, it is virtually impossible, and impractical, to provide a jettisoning device that would not impart a tumbling moment to the manipulator that could cause it to impact and possibly damage the fragile thermal protection system that covers the orbiter's skin. The solution is to activate the reaction control system and displace the orbiter away from the manipulator (and pedestals) the instant the retractors have fired.

Figure 6-18 also shows the retractor very much resembling a thruster that has already functioned. A shear pin (not shown in the illustration) locks the piston in its extended position. Note how the orifice from the cartridge plenum ports the high-pressure gas to the shank side of the piston. Here the gas pressure pushes on an annulus defined by the piston and the shank outside diameters. Like the thruster, first motion will shear the locking pin. When the piston moves the entire length of the housing, a thin-walled skirt machined into the piston strikes a beveled anvil built into the housing cap, causing the skirt to flare into a recess in the base of the cap. This flaring skirt locks the piston in the retracted position, preventing it, and the toggle mechanism, from rebounding and possibly jeopardizing a successful separation from the orbiter. The extended end of the shank is threaded to accept a clevis fitting, with a matching thread, that attaches the retractor to the toggle mechanism. Only the initial 0.06 inch (1.5 millimeters) of retraction experiences high loads, while the remainder of the stroke is free travel to accommodate the kinematics of the separation toggle mechanism. These can be a combination as high as 200 to 700 pounds (91 to 318 kilograms) side and axial loads respectively. In addition to these resisting loads the locking shear pin must be broken. For a performance margin demonstration, the reactor must function with the above loads applied, utilizing a single pyrotechnic cartridge containing a prescribed downloaded (reduced) main propellant charge.

Like the thruster previously described, this illustrated retractor incorporates only a limited number of the capabilities and functions that retractors can perform. Metering pins, honeycomb shock attenuators, etc., that were described for thrusters can be adapted to retractors. The variations are limited only by the experience and imagination of the pyrotechnic designer.

## Pin-Pullers

While preparing for the Apollo lunar missions, scientists and engineers had a vital need of closeup photographs of the moon's surface. Such photographs would be necessary in the design of future landing craft. The spacecraft built to fulfill this requirement was called Ranger. During its final phase of the mission its transmitting cameras, aligned with its longitudinal axis, were turned on to rapidly transmit photographs back to earth. The ninth and final Ranger transmitted 5,814 photographs, with the final frame achieving lunar feature resolution approaching one foot (0.3 meter). Figure 6-19 illustrates the spacecraft consisting of a hexagonal base structure, 5 feet (1.5 meters) in diameter, with a large parabolic antenna extending from beneath and a tower structure built atop the base containing radio transmitters, sensors, stabilization jets, experiments, etc. The overall height was 11 feet (3.5 meters) and weighed 675 pounds (306 kilograms). The Ranger was the first spacecraft to use deployable solar arrays. There were two solar panels, each approximately 1.5 by 6 feet (0.5 by 1.8 meters), extending from opposite sides of the spacecraft. They were hinged to the lower portion of the base structure and during launch were folded upward toward the tower structure. About halfway up the tower, adjacent to the folded panels, were attached a pair of pyrotechnic pin-pullers. Their extended pins engaged a clip attached to each of the solar panels and held them in the folded position. Approximately a half-hour after launch a sequencer automatically fired the pyrotechnic cartridges, with battery power, pulling the pins out of the clips and, with the aid of cocked springs in the panel hinges, causing the solar panels to rotate approximately 90 degrees. Sensors then fired the gas jets that rotated the spacecraft in the proper direction to orient the solar panels toward the sun. Once oriented, the batteries were turned off and the solar arrays supplied all the spacecraft's electrical power—up to 210 watts.

Each solar panel pin-puller weighed 2.4 ounces (68 grams) including cartridges. The pin diameter is 0.186 inch (4.7 millimeters) and retracts 0.5 inch (12.5 millimeters) with a 3.3 foot pounds (0.43 kilogram meters) bending

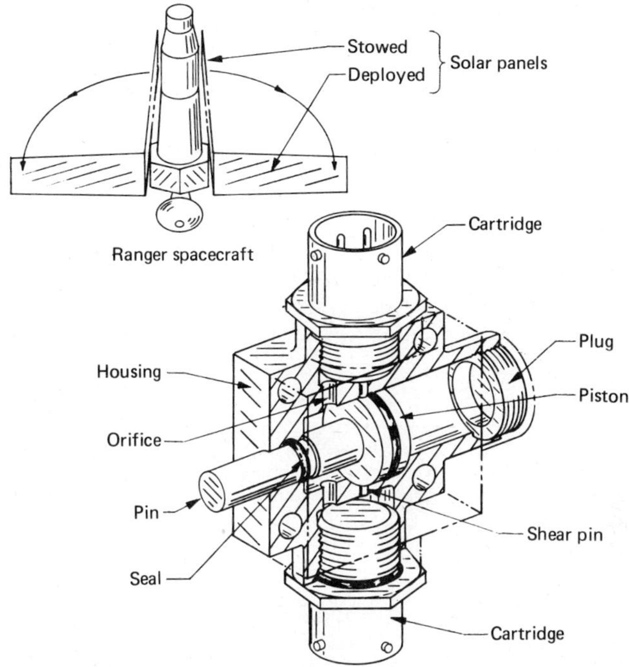


**FIG. 6-19.** Ranger spacecraft solar panel deployment pin-puller.

moment on it. Like the retractor and the thruster previously discussed, the piston is retained in its initial position by a shear pin that must be broken during its initial movement. Like the retractor, the high-pressure gas generated by the pyrotechnic cartridge is ported to the net side of the piston, causing the extended pin (shaft) to retract into the housing. Only the gas pressure and the friction of the two seals prohibit the piston from rebounding after impacting the closeout plug at the end of the stroke.

## SWITCHES AND VALVES

Cartridge-actuated switches and valves are a group of fast-acting pyrotechnic devices that can function as an emergency safety device or as a routine operation at a specific time in a sequence. Both devices can be designed to perform many different functions, with the primary differences being that switches are generally associated with electrical systems while valves are components of hydraulic and pneumatic systems. Both can be normally closed or normally open, or can change the routing of the energy medium from one circuit to another. The following examples illustrate pyro-techniques that build upon the basic principles of retractors and thrusters to create a new group of devices performing totally different functions.

## Transformer Protective Switch (TPS)

An electric utility decided to convert its subtransmission system from 69 kilovolts (kV) to 138 kV due to a rapid load growth and increasing load density in one of its metropolitan areas. Existing transformer protective power fuses could not be used since they lacked sufficient power interrupting capability.

Figure 6-20*a* is a photograph of a transformer protective switch as it would be installed on a pole unit. Figure 6-20*b* is a schematic of its major components. $SF_6$ (sulfur hexafluoride) is a high dielectric gas that is caused to flow across the opening contact, by the moving interrupter, thereby reducing the potential (of arcing) in the initial stage of current interruption.

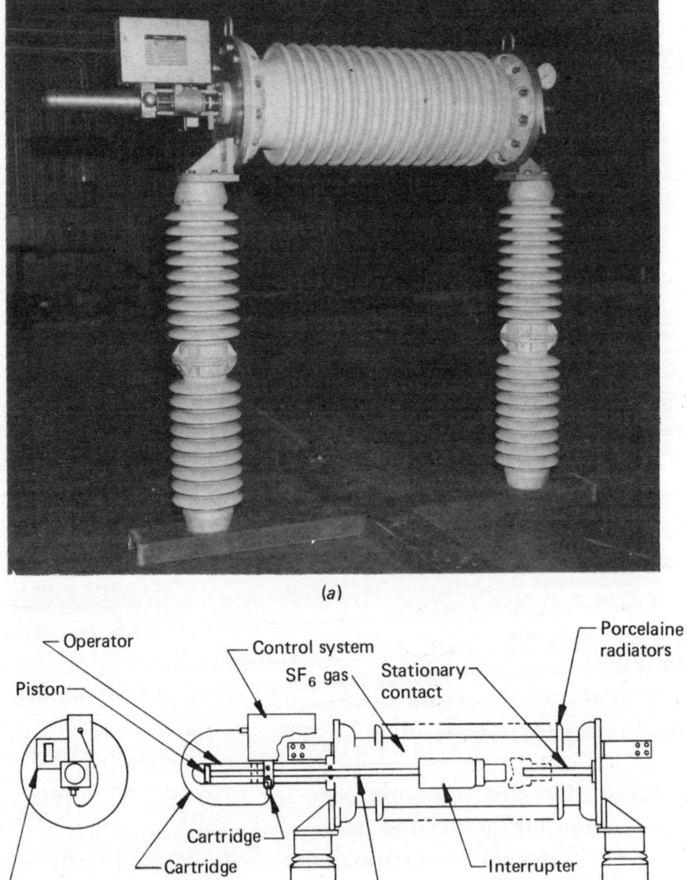


**FIG. 6-20.** Transformer protective switch (TPS). (*a*) TPS external view. (*b*) TPS schematic diagram.

The control system contains a current sensor and other components that provide both short delay response of approximately 1 second, for short-duration, low-current surges, and instantaneous responses for high over-current surges and direct shorts. The response function of the control system is to send an electrical discharge pulse to the pyrotechnic cartridge. Like the retractors, the gas is ported to the net side of the piston whose connecting rod is attached to the interrupter which is mated to a stationary contact. The high-pressure gas causes the piston, rod, and interrupter assembly to move away from the stationary contact a sufficient distance, approximately 8 inches (203 millimeters), to interrupt the current flow.

The transformer protective switch will not require maintenance under normal service. Its operating life is 50 cycles over a 45-year time span with the necessity of the operator being properly cleaned each time the cartridge is fired along with installation of a new cartridge. The contact life of the interrupter is the integrated sum of 450-kiloampere (kA) operations or, for example, fifteen 30-kA fault operations.

## Explosive Valves

The adaptation of the gas-generating pressure cartridge to remotely operated one-shot valves to control the flow of liquids or gases was one of the earliest applications of pyrotechnic energy.

Two basic types of valves have evolved, "normally open" and "normally closed." These names identify the prefunction characteristic of the valve. An example of each valve type is illustrated in Figure 6-21.

**Normally Open Explosive Valve** The normally open valve (Figure 6-21*a*) consists of a housing with an inlet and an outlet port with no obstructions to prevent the free flow of fluid or gas through the valve. Axially aligned with the outlet port is a piston with a pyrotechnic pressure cartridge installed behind it. On the opposite end of the piston is a shallow-angle conical surface. An almost matching female cone is machined into the housing at the entrance to the outlet port. Upon initiation of the pyrotechnic cartridge, high-pressure gas drives the piston forward at a high rate of acceleration. When the piston cone contacts the nearly matching outlet port entrance cone under these conditions, the contacting surfaces deform, creating a highly efficient, metal-to-metal seal. This seal immediately stops the flow of liquid or gas within 5 milliseconds.

Normally open explosive valves have many aerospace applications to stop the flow of liquid propellants in rockets in preparation for stage separation since it is not known exactly when all of the propellant will be consumed. By stopping the propellant flow with an explosive valve, this event can be precisely controlled. On high-altitude rockets, this requirement becomes even more complicated because the propellant actually consists of two separate fluids: a fuel and an oxidizer. The flow of each must be terminated instantaneously to avert a hazardous condition within the rocket engine. Fast-acting, normally open explosive valves are a natural for this application.

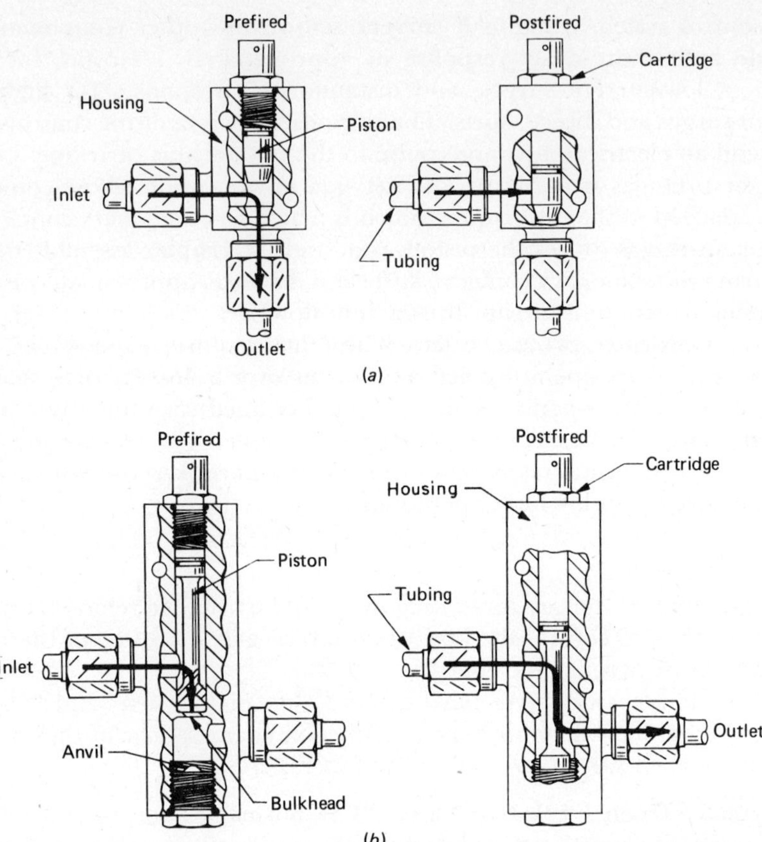


**FIG. 6-21.** Explosive valve. (*a*) Normally open valve. (*b*) Normally closed valve.

**Normally Closed Explosive Valve** The normally closed valve (Figure 6-21*b*) has an integral bulkhead machined into the housing that is also axially aligned with the pyrotechnically driven piston. On the end of the piston that contacts the bulkhead is a cutting edge similar to that of a punch press cutter. As the piston severs the bulkhead, it travels by the outlet port until it impacts an anvil machined on a closeout plug at the end of the transfer orifice. The purpose of the anvil is to deform the severed bulkhead in such a way as to lock the piston to the closeout plug. If this feature were not provided, the fluid or gas pressure on the under side, or net side, of the piston could position it in an intermediate position, resulting in partial flow restriction to the outlet port. Note how the reduction in diameter of the piston rod connecting the piston to the bulkhead cutter allows the fluid or gas to travel around the rod from the inlet to the outlet ports. To prohibit flow restrictions, the area of the annulus of the transfer orifice around the piston rod must be equal to or greater than the area of the inlet and outlet ports.

Normally closed explosive valves are used on remotely located fire extinguishers and are controlled by heat- and smoke-sensitive switches. They have been used in aircraft fuel systems to "dump" excess fuel in preparation for an emergency landing, and to release high-pressure gas from a storage bottle into a bladder of a hydraulic accumulator, thus pressurizing the hydraulic system to perform useful work such as an emergency control system on an aircraft or ship.

These two valve concepts can be partially combined to provide numerous switching functions such as normally open, normally closed switch or normally closed, normally open switch. These valves require housings with common inlets and multiple outlets. When the valve is functioned, an outlet that was initially closed is opened and an outlet that was initially opened is closed.

## ELECTRIC UTILITY PRODUCTS

The installation and maintenance of electric power transmission and distribution lines has recently felt the inroads of powder-actuated tools and devices. These include taps (or connections) into main and secondary transmission lines with power in the through conductor either off or on. These taps are made with installation tools that are powder-actuated. In addition, conductor splices, dead ends, terminals, and lugs can now be installed by internally fired charges that are initiated by a small lantern battery or by a sharp blow with a hammer. Since most of this work is done atop power poles or towers and in deep, narrow ditches, it can be readily understood how portable powder power could simplify the job when compared to the conventional method of employing pneumatically or hydraulically powered tools to accomplish the same functions. The cost of hardware is comparable for like types of connections, e.g., splices, taps, and terminals. The biggest cost savings for the utility company are realized by selecting the type of hardware that requires both the least expensive installation equipment and the least number of worker-hours to operate it. The reader can imagine the time it would require to move a trailer containing hydraulic or pneumatic power units, distribute hoses, and climb poles or towers each time a piece of hardware was installed. Examples of several types of powder-actuated electric utility power transmission devices are discussed in the following pages that will once again enable the reader to appreciate how the precision application of powder-actuation has simplified many heretofore complicated, time-consuming, expensive, and oftentimes risky tasks.

### Line Taps

A line tap is a connection into a main transmission line (through line) away from a terminal. This can occur at mid-span, near a terminal, or anywhere in between. Figure 6-22 shows two views of a tap with the through line at the left in each view. In addition, there is the tap wire that begins at this location

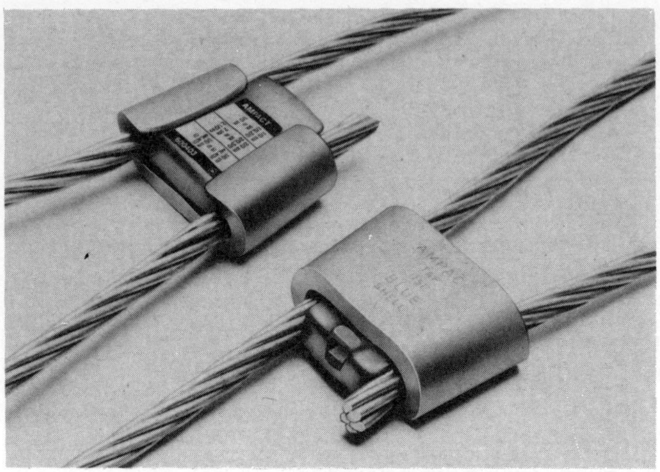

**FIG. 6-22.** Power transmission line tap. (Ampact is a trademark of AMP Incorporated.)

and the two mechanical components that firmly support both the tap wire and the through line. Both tap components are wedge-shaped when viewed in a plane parallel to the connection. In cross section they are quite different. The outer C member retains both the through line and the tap wire firmly against the inner wedge that has been powder driven between the two wires. Materials for both components can be either aluminum or copper. Both components have been coated with an inhibitor, containing abrasive particles, over the common surfaces that contact the two wires. The purpose of the abrasive is to remove any oxidation from the contact surfaces between the wires and the tap components as the wedge is being driven into the C member. The contact area between the tap and the through line and the tap wire is greater than the cross-sectional areas of both wires.

The installation of a tap is quite simple and begins by orienting the C member with the open side away from the lineman as shown in Figure 6-23*a*. Whether the wider end of the C member is to the right or left is dependent upon which side the tap wire enters the tap. The best connection results when the tap wire enters the tap via the widest end of the tapered C member. In the example, the tap wire enters from the right so the widest end of the C member is oriented to the right. A mating wedge is aligned with its narrow end entering the widest opening of the C member. The wedge is pushed, by hand, forward inside the C member and between the through line and the tap wire. A few light taps with a hammer will drive the wedge far enough to hold the tap wire to the through line firmly enough while the tap installation tool is being positioned around the tap.

In appearance, the tap application tool grossly resembles a C clamp. That is, it has two components threaded together. The C shaped component containing the female thread at one end and an anvil-like surface at the

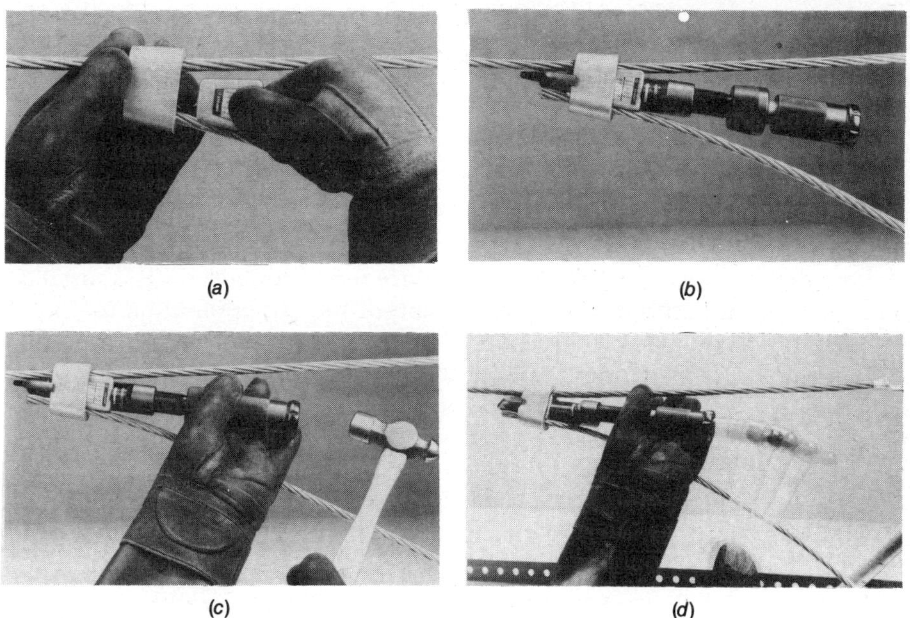


**FIG. 6-23.** Line tap installation. (*a*) Positioning the C member and wedge. (*b*) Installing the tap application tool. (*c*) Verifying tap application tool striking clearance. (*d*) Firing the tap application tool.

other is called the tool head. The male threaded shank assembly containing the pyrotechnic cartridge is called the power unit. The power unit is backed out of the tool head far enough to permit the anvil portion of the tool head to be placed against the narrow end of the tap's C member as shown in Figure 6-23*b*. Note the closed side of the tool head is on the open side of the C member. The power unit is then threaded firmly against the wide end of the tap's wedge. This places the power unit midway between the through line and the tap wire. The lineman then supports the tap application tool, by the power unit, with one hand and aligns a hammer held in the other hand (as shown in Figure 6-23*c*), with the free end of the power unit. The lineman is verifying there is adequate clearance to strike the end of the power unit with the hammer with sufficient force to initiate the powder cartridge within.

When the lineman is assured he has adequate clearance to initiate the tap application tool, he firmly strikes the free end of the power unit with the hammer. This action initiates an internal gas-generating powder cartridge that drives a ram within the power unit firmly against the wedge of the tap. Figure 6-23*d* shows the wedge driven well within the C member. In preparation to remove the application tool from the installed tap the lineman backs off the gas pressure release knob from the end of the power unit he had just struck with the hammer. When all the excess gas pressure has been vented from the power unit, the power unit can be unthreaded from the tool head.

Once adequate clearance has been established between the power unit and the tool head, the tap application tool can be easily removed from the installed tap.

The tap locking forces are both friction and mechanical deformation between the C member, wedge, through line, and tap wire. The C member is designed to spring-open slightly during wedge insertion, to maintain a constant minimum squeeze force on the tap during its subsequent exposure to radical temperature changes in the local environment. Figure 6-22 explained the purpose of a line tap and Figure 6-23 showed how it is installed on a through line in the field. A discussion of the tap application tool itself will further clarify its internal sequence of events.

Figure 6-24*a* is a cutaway drawing of a line tap application tool and the C member and the wedge of the tap. There are only two moving parts within the tool, the ram and the gas check. In preparing the tool for firing, the breech cap and gas release knob are removed, as an assembly, from the power unit. A tapered powder cartridge containing a plastic shell, the gas generating charge, percussion primer, and molded gas check are slipped over the aft end of a cylindrical ram with a built in firing pin on its aft end. The tapered shell of the cartridge extends well along the shank of the ram. The breech cap and gas release knob assembly are threaded tight against the

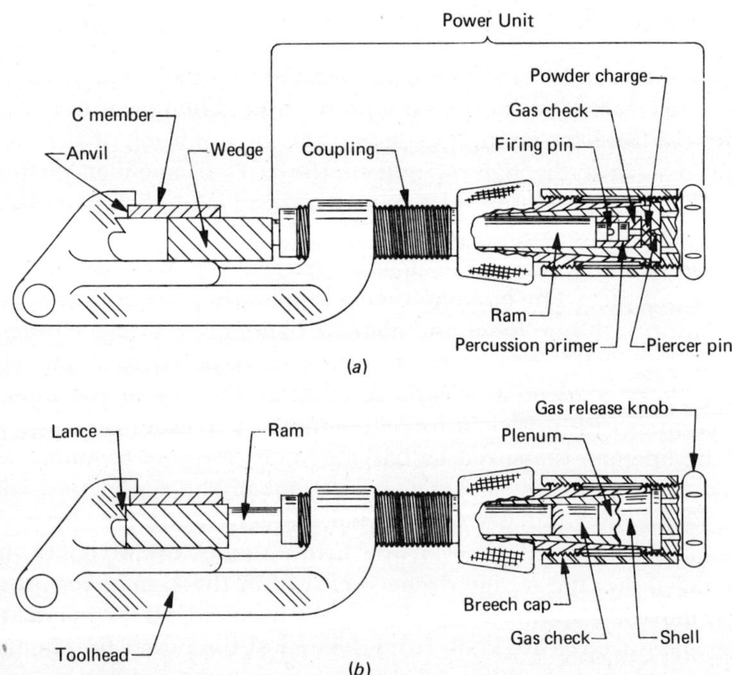


**FIG. 6-24.** Cutaway of line tap application tool. (*a*) Tool ready for firing. (*b*) Fired tool.

cartridge. During the final turns a tapered piercer pin protruding forward from the center of the gas release knob pierces the cartridge chamber containing the powder charge. The forward end of the gas check is composed of several legs (safety prongs) that prevent the firing pin on the end of the ram from contacting the percussion primer within the gas check. During the act of the aforementioned piercer pin penetrating the cartridge shell, the safety prongs are being crushed, allowing the gas check to move toward the firing pin, but still not allowing the percussion primer to contact the firing pin on the end of the ram. This is the internal condition of the tap application tool as shown in Figure 6-23*c*. When the power unit is abruptly hit with a hammer on the gas release knob, the free sliding ram moves relatively aft with enough force to further crush the safety prongs extending from the gas check, thus allowing the ram's firing pin to strike and initiate the percussion primer (see Chapter 3). The flame and shock wave from the primer travel through a "spit" hole within the gas check to the powder charge beyond. Immediately, the gas pressure generated by the powder charge causes the gas check and ram to move forward with tremendous force, as shown in Figure 6-24*b*. The opposite end of the ram had been previously positioned against the tap wedge. As the wedge moves forward, the C member is prevented from likewise moving by the anvil on the tool head.

A second feature on the tool head, near the anvil, is called a "lance." It serves the dual purpose of preventing the tap wedge from being driven too far within the C member and it also deforms the contacting edge of the wedge upward as it exits the edge of the C member as is clearly evident in the lower view of Figure 6-22. This deformation provides a third method of locking the wedge within the C member in addition to the two previously described. Even though the ram has stopped moving, gas pressure still exists behind the ram. If the power unit was unthreaded from the tool head the ram would continue to extend, causing an unsafe situation for the lineman. To alleviate this condition, prior to unthreading the power unit from the tool head, the gas release knob is unthreaded several turns first. This action withdraws the piercer pin from the powder charge chamber of the cartridge allowing the excess gas to escape through vent holes in the knob. When completely vented, the power unit can then be safely unthreaded from the tool head to permit removal of the tap application tool. The breech cap and gas release knob assembly are then removed and the spent cartridge is removed from the end of the ram. Then the sequence is ready to begin again when the next transmission line tap is to be installed.

Close scrutiny of Figure 6-23 indicates there was no electric current in either the through line or the tap wire during the tap installation. A tap can, however, be made to a "hot" line by slightly modifying the procedure just described. Instead of grasping the hot line with his hands, the lineman uses a "hot-stick" kit that consists of a group of tools, i.e., holders, hammers, and clamps mounted on the end of short dielectric sticks that actually contact the hot line, tap wire, C member, wedge, and application tool. It is surprising how, after a little practice, linemen can become quite proficient with the hot-stick method of installing line taps.

Whenever a tap is made to a hot line, arcing often occurs that results in slight damage to the arcing components. If the tap is made to the hot line only once at a particular location, the arcing damage can be tolerated. If, however, the tap is to be made and removed several times, then the cumulative effect is undesirable and the hot line can be severely damaged. Figure 6-25 shows an installed tap stirrup that eliminates this condition. The stirrup is tapped to the hot line, with no arcing because the tap circuit (the stirrup) remains open. The tap wire is then attached, utilizing the hot-stick method, to the stirrup where the arcing occurs, leaving the hot line undamaged. After several removals and reinstallations of the tap to the stirrup renders the stirrup unusable, it can be removed and/or replaced by another stirrup.

The line tap method has also been adapted to terminal installation at the end of transmission lines. Figure 6-26 shows one such terminal from a broad range of terminal configurations available. They provide the capability of being able to disconnect the line by simply removing a number of nuts and bolts, thereby leaving the line/terminal interface unscathed.

When a tap to a transmission line has to be removed, the same tap application tool can be fitted with an adapter that mounts on the frame of the tool head. The tap application tool is installed on the opposite side of the tap as was previously described. The tool-head adapter elevates the tap such that the ram reacts against the C member and the anvil reacts against the

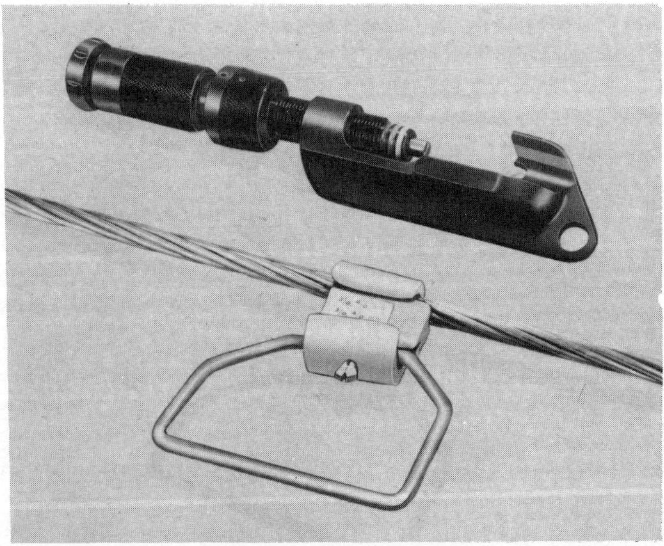

**FIG. 6-25.** Line tap stirrup.

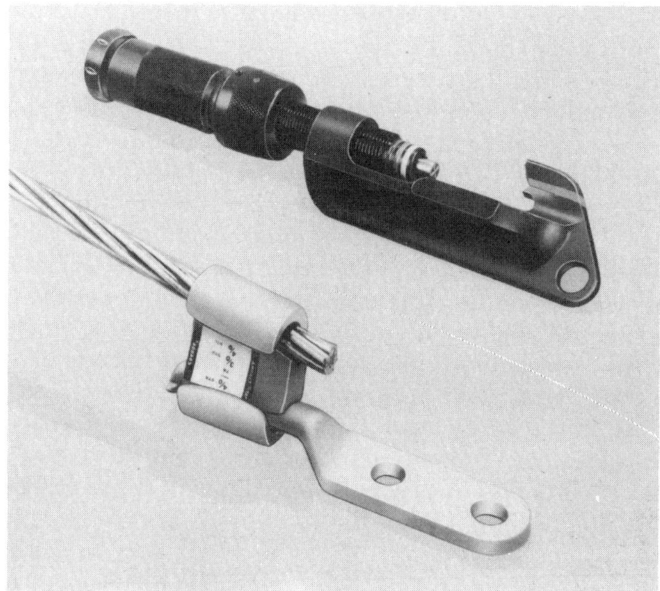

**FIG. 6-26.** Line tap terminal.

wedge (just the reverse as was described for installing the tap). Then the firing procedure is the same as when the tap was installed.

Each tap for every through line and tap wire size and material (copper and/or aluminum) combination has been analyzed, resulting in the necessity of four-power-level powder cartridges to cover the entire range. Although they are the same geometric size, the cartridges are color-coded red, white, blue, and yellow to indicate the order of increasing power levels. Each tap C member and wedge has stamped on it the proper cartridge color to be used in the application tool for tap installation. The same cartridge information is also available for removing a tap. The most powerful cartridge, yellow, is never used to remove a tap and it requires a special, stronger tool head to install the larger taps.

## Internally Fired Connectors

Even though the powder charge in each of the following examples is not a separable cartridge, as is indicated by the title of this chapter, they are included here as part of the general heading of "Electric Utility Products."

The powder charge of the previously described line taps was located within the application tool. There are also many transmission line connectors having nonseparable, internal powder charges that supply the energy for their installation. Two methods of initiating the charges within, e.g., percussion and electrical, are available. Selected examples of both types are reviewed.

**Percussion-Fired Lugs** Perhaps the simplest method of all for installing lugs or terminals on the ends of transmission lines utilizes the internally fired, percussion-initiated method of installation. Figure 6-27*a* shows a transmission line with the insulation removed a prescribed distance from the end of the line. The bare end of the line is inserted into the hollow shank of the lug. The lineman then removes the safety pin, Figure 6-27*b*, from the lug shank and grasps the line near the shank of the lug with one hand. With his other hand, the lineman takes a hammer and firmly strikes the end of the lug as demonstrated in Figure 6-27*c*. The blow initiates the internal percussion primer that fires the main powder charge resulting in permanently attaching the lug to the transmission line.

The internal functions of the lug are similar to those of the line-tap application tool. Figure 6-28, from top to bottom, illustrates the sequence of events that occurs within the lug during installation. The internal components consist of a powder charge in a blind hole within the lug. Plugging the powder hole is a percussion primer followed by a gas seal (or gas check). A piston ram with a firing pin built into its shank is held in place with a safety pin. Next to the piston is a four-segment set of jaws with tapered outer surfaces that matches the internal tapered shank of the lug. The internal surfaces of the jaw segments are grooved to better grip the transmission line. They are also coated with an inhibitor containing abrasive particles to remove oxidation from the line strands during installation. After the safety pin has been removed and the end of the lug is struck with a hammer, the piston abruptly moves toward the percussion primer, causing the integral firing pin to initiate the primer that in turn fires the main powder charge. The gas pressure then reverses the action by driving the gas seal and piston against the segmented jaws. As the jaws move aft, the matching tapered surfaces of the segments and the shank cause the segments to be squeezed against the exposed transmission line within the segmented jaws. Friction between the tapered surfaces of the segments and the shank is sufficient to prevent the residual forces from the compacted transmission line to cause the segments to slide in the opposite direction after the gas pressure behind the piston has dissipated. An extremely small angle of taper also enhances the segment locking characteristics.

**Battery-Fired Connectors** Oftentimes it is necessary to splice a transmission line by inserting two line ends into a common splice connector. Some splice connectors have a special requirement in addition to maintaining electrical continuity from the donor line to the acceptor line and that is to provide mechanical strength equal to or greater than that of the individual lines being spliced. The same requirement often befalls transmission line terminal connectors. Conversely, a splice or terminal connector may require very little mechanical strength, compared to that of the transmission line. Figure 6-29 presents two examples of both types of connectors. From left to right, the first two are a splice and a terminal connector that must provide,

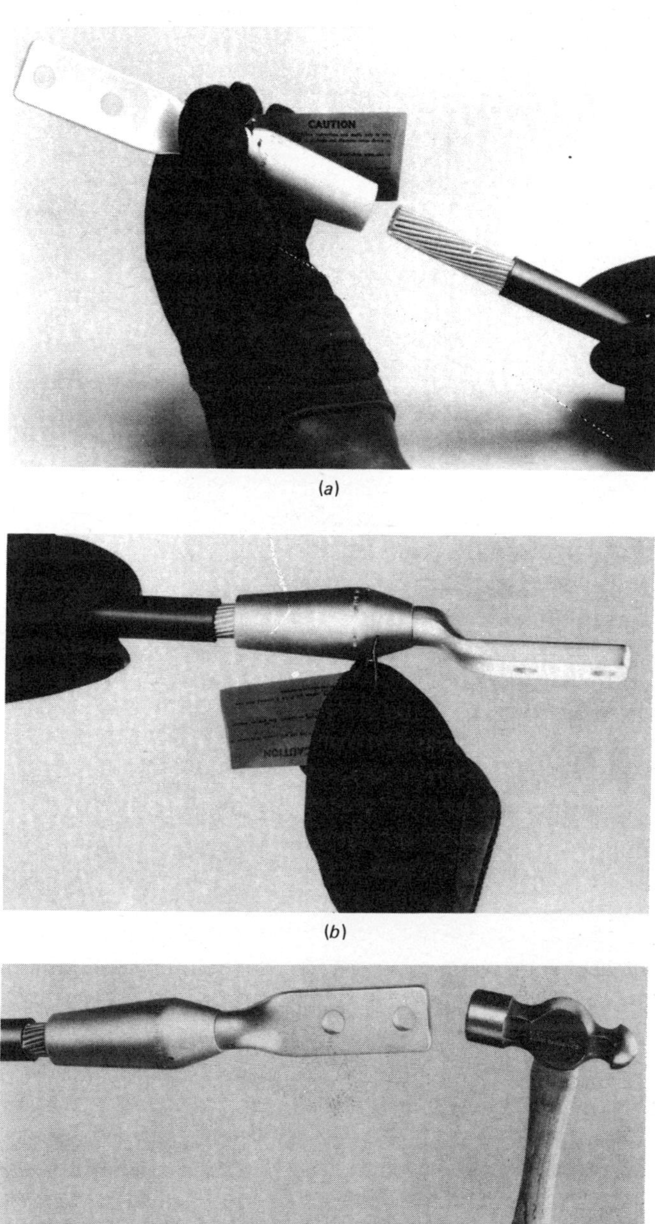


**FIG. 6-27.** Percussion-fired lug installation. (*a*) Line insertion into lug. (*b*) Removing the safety pin. (*c*) Initiating the lug's internal powder charge.

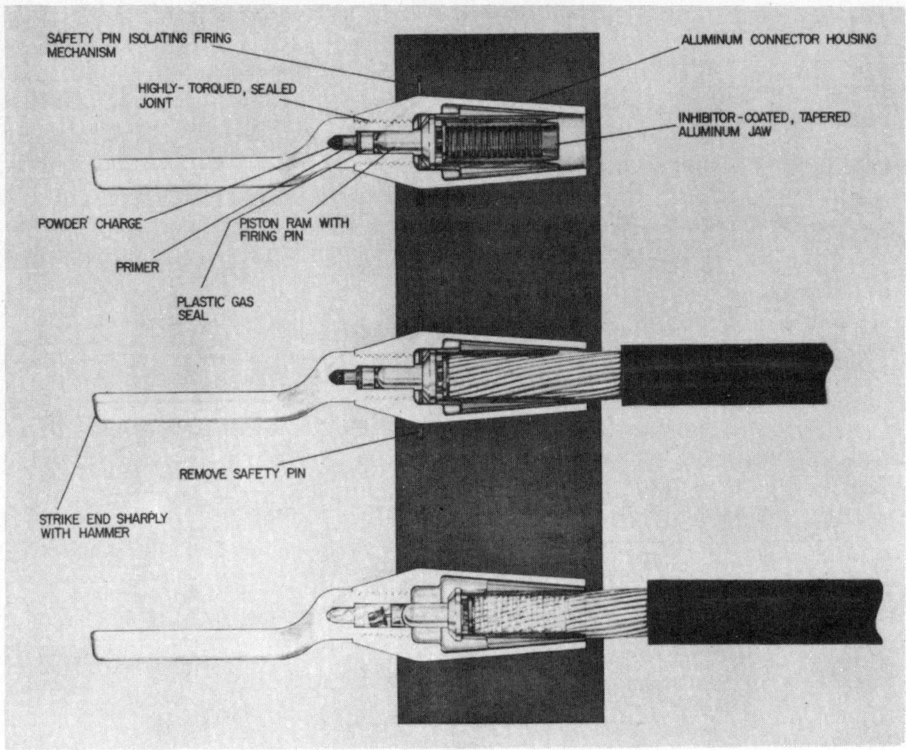


**FIG. 6-28.** Cutaway of percussion-fired lug.

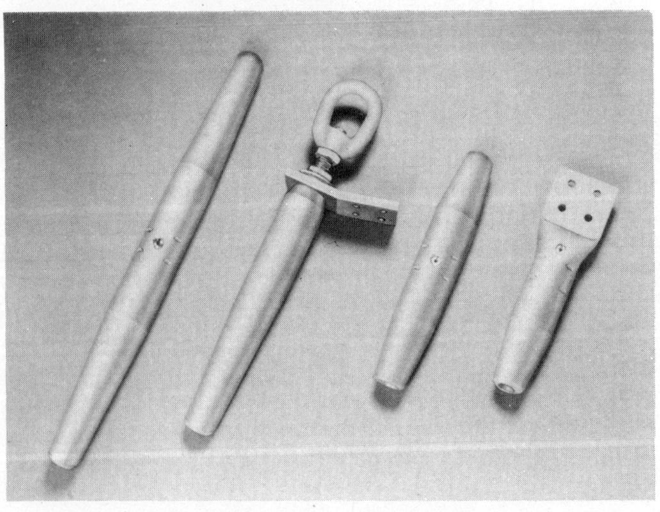

**FIG. 6-29.** Battery-fired connectors.

after installation, mechanical strength equal to or greater than the transmission line onto which it is installed. In electric utility jargon, these are Class 1 connectors. The splice and terminal connectors on the right side of the photograph are Class 3 connectors, that is, their mechanical attachment to their respective transmission lines are not necessarily equal to that of the line. The two main characteristics they all have in common are that they all utilize a 6-volt lantern battery to initiate the internal powder charge and that they all utilize segmented jaws to squeeze against the transmission lines to which they are attached. An examination of the internal configurations of the two classes of attachments will clarify how one has considerably more mechanical attachment strength to the transmission line than does the other.

*Class 1 connectors* Electric utility transmission lines are constructed of two material types of strands within the same line. The core strands, which total about half the diameter of the line, are made of steel, whereas the outer strands are of aluminum. They all transfer electric power. On large diameter lines over long spans between transmission support towers, an all aluminum line would not adequately support the massive weight of the line. The two-conductor line is a compromise that employs the most desirable properties of aluminum for light weight and current transmission capabilities and steel for superior mechanical strength. Figure 6-30 is a cutaway drawing of the Class 1 splice connector on the left of Figure 6-29. From top to bottom is illustrated the splice connector by itself, in the middle with transmission lines inserted and ready to fire and on the bottom the splice connector simultaneously installed on two transmission lines. On the left of the upper view will

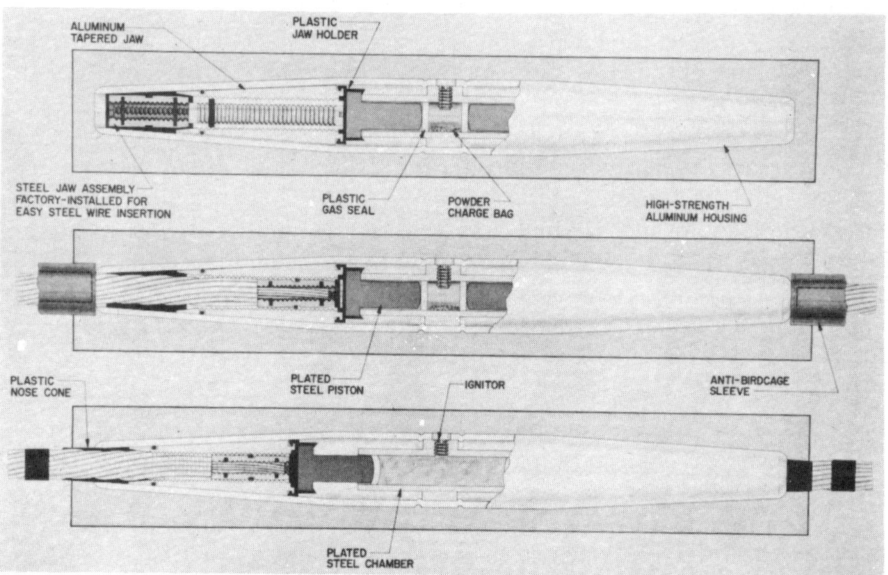


**FIG. 6-30.** Cutaway of class 1 splice connector.

be noted there are two sets of segmented, grooved jaws. A small diameter set of steel jaws to grip the inner steel strands of the transmission line and a larger diameter set of aluminum jaws to grip the outer aluminum transmission line strands. In preparing the transmission line for insertion into the splice connector, the outer aluminum strands are removed for a distance slightly less than the length of the inner steel jaws. As the line is inserted into the splice connector the exposed small diameter steel core strands engage the steel jaw segments located at the entry of the splice connector. When the larger diameter shoulder of the outer aluminum strands contacts the steel jaw segments, they are displaced axially toward the center of the splice connector and inside the outer aluminum jaws. Note the aluminum jaws are approximately twice as long as the inner steel jaws. The outer diameter of the aluminum jaws matches a tapered housing surface as did the percussion-fired lug of Figure 6-28. After both transmission lines have been inserted into the splice connector, an anti-birdcage sleeve is wrapped around each line adjacent to each end of the splice connector. This sleeve prevents the line strands from bulging during the firing operation.

Dual pistons at each inner end of the segmented jaws have their shanks near a powder chamber at the center of the splice connector. An igniter connects the powder chamber to a detachable firing collar (not shown in the illustration). When inserted in the igniter and attached to a 6-volt lantern battery, the firing collar initiates a small quantity of priming mix at the base of the igniter located at the entry into the powder chamber. The priming mix initiates the powder charge that generates high-pressure gas to drive the two pistons in opposite directions. The pistons displace the four sets of segmented jaws (two steel and two aluminum) and two transmission lines axially along the center line of the splice connector. In doing so, the outer tapered surface of the aluminum jaws are squeezed by the matching taper of the housing. The outer aluminum jaws squeeze directly against the outer aluminum strands of the transmisison line. At the piston ends of the aluminum jaws, they squeeze onto the inner steel jaws that in turn squeeze onto the inner steel strands of the transmission line. Together the mechanical-friction attachment of the dual jaws results in a joint that is as strong or stronger than the breaking strength of the transmission lines themselves. A proper splice connector installation is indicated by the exposure, at either end of the connector, of a plastic nose cone (sleeve) that was displaced with the transmission line and dual sets of jaws during the firing sequence.

*Class 3 connectors* Whenever a splice or terminal does not have to provide the mechanical strength of the transmission line, a Class 3 connector is selected. The splice and terminal connectors on the right side of Figure 6-29 are Class 3 connectors. The single difference between the Class 1 connector of Figure 6-30 and the Class 3 connector of Figure 6-31 is almost immediately discernible to be the Class 3 connector has only one set of jaws (aluminum) squeezing the transmission line. The internal configuration is very similar to

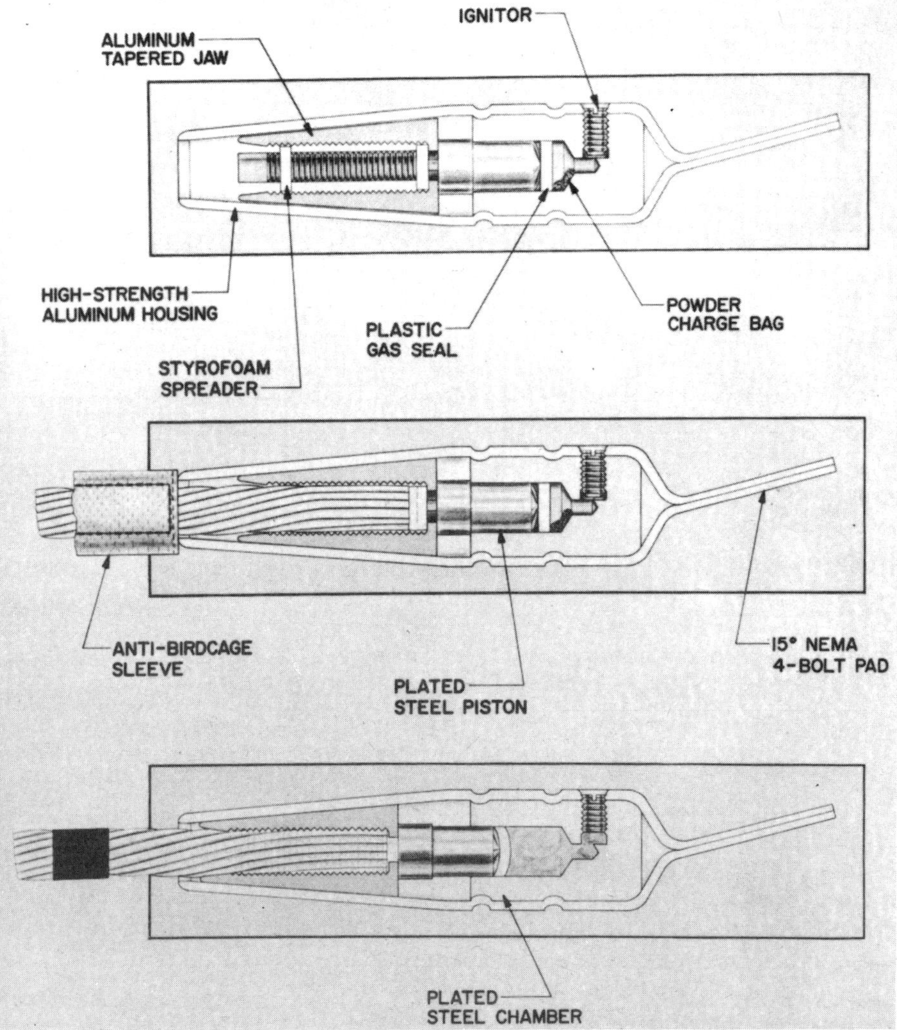


**FIG. 6-31.** Cutaway of class 3 pad connector.

the percussion-fired lug connector described in Figure 6-29. The Class 3 connector employs the same firing collar that attaches to the connector igniter and the 6-volt lantern battery.

*Underground distribution splice* More and more communities are requiring that utility distribution lines be buried underground. In addition to being more time-consuming and expensive to install, special environmental requirements are placed on the splicing of these transmission lines. Figure 6-32 is a schematic of one type of underground distribution splice. To accommodate

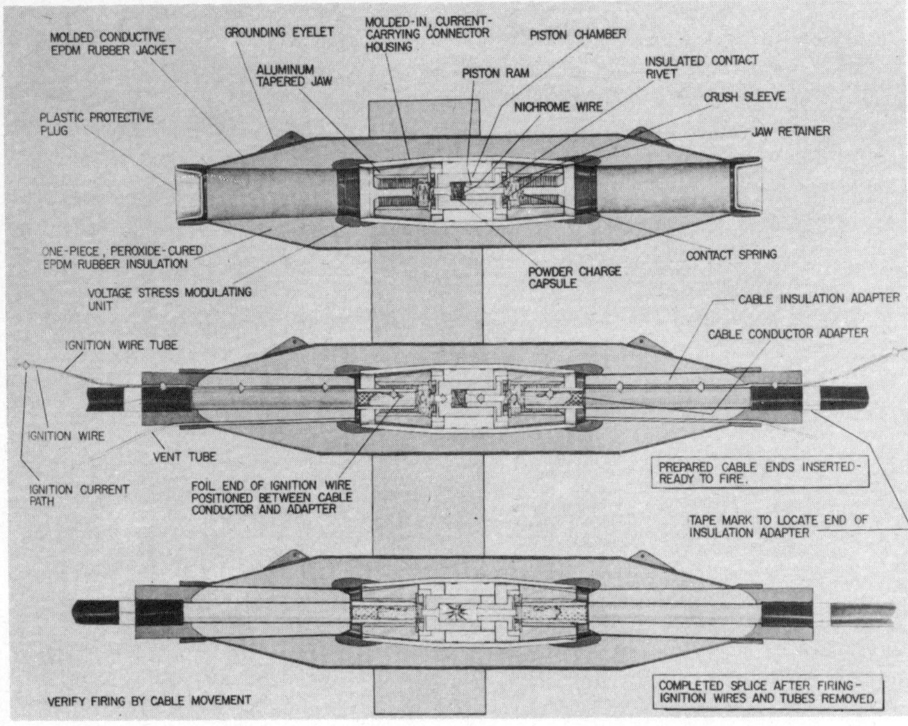


**FIG. 6-32.** Cutaway of underground distribution splice.

the many different sizes of transmission lines and at the same time minimize cost and warehousing requirements, there are two sets of concentric segmented jaws squeezing each transmission line. To accommodate different-diameter transmission lines, a specific set of inner conductor adapters (grooved, segmented jaws) must be used. All conductor adapters have the same outside diameter, the grooved inside diameters vary to accommodate the varying line diameters. As in the previously described internally fired connectors, the center powder charge is fired with a 6-volt lantern battery that drives opposing pistons against segmented jaws against a common tapered cone. The taper squeezes the outer jaws against the connector adapters that in turn squeeze the transmission line. The wires from the battery enter from either end of the splice connector as far as the conductor adapters. From there the electrical path is through the transmission line a short distance, then through a compression spring to an insulated contact rivet that contacts the metal capsule at the center of the splice that contains the powder charge. A bridgewire through the capsule is heated by the battery current and initiates the powder charge. The vent tube provides an escape path for the trapped air as the insulated transmission lines are inserted into the connector. Both these vent tubes and ignition wire tubes are removed after the powder charge has been fired. It is evident the remaining differences are to accommodate the varied underground environments encountered, i.e., hermetic seals and thermal insulation. The underground distribution splice does not have to meet the mechanical strength requirements of a Class 1 connector since the weight of the transmission lines are continuously supported by the bottom of the ditch.

chapter 7

# Specialized Pyrotechnic Devices and Systems

There frequently is an application of a pyrotechnic device or system to meet very unique requirements. Indeed, many of the pyrotechnic applications of the previous chapters could have initially met this definition, but as their application became widespread, they became commonplace. This chapter discusses examples of these specialized pyrotechnic devices and systems that have been selected from the broadest base available.

## SAFE-AND-ARM DEVICES

Safe-and-arm (S & A) devices are placed in the initiating systems of many aerospace pyrotechnic systems such as shroud separation, stage separation, destruct systems, propellant dispersal systems, and thrust termination systems. Their most widespread applications, however, are in rocket motor ignition systems and especially solid rocket motors (SRM). The word "solid" indicates that the propellant within the rocket is a solid as opposed to liquids such as those used in the three stages of the Apollo Saturn V rockets. The two primary purposes of the S & A device are to inhibit premature ignition of the SRM in the SAFE position and then reliably initiate the SRM when their pyrotechnics are fired in the ARM position. There are probably more variations of S & A devices than any other category of pyrotechnic device. This is due primarily to the hundreds of SRM configurations that have been in existence. Although all S & A devices serve the same primary purposes described above, each has been designed to accomodate a specific SRM with its own peculiar requirements.

S & A devices are located at the forward end of SRMs and must have the capability of being SAFED or ARMED, at will, by the launch controller from a remote location. In addition to pyrotechnic components, they also require electric motors, switches, position indicators, mechanical linkages, gears, pyrotechnic barriers, and many other components, depending on the requirements of the applicable SRM. As one might suspect, with so many S & A device requirements to fulfill and numerous options to meeting each requirement, there are many configurations available that fulfill specific S & A device requirements. With this in mind the following S & A device requirements were excerpted from the initial space shuttle solid rocket booster ignition requirements specification. (A solid rocket booster, SRB, is an SRM plus all the additional systems such as nozzle thrust, vector control, parachute recovery, destruct system, etc.)

The space shuttle's two SRBs are approximately 12 feet (3.6 meters) in diameter and 120 feet (36.5 meters) long, and during launch they continuously burn along their entire length for over 120 seconds. Their thrust output is over 2.5 million pounds (1.13 million kilograms) each and is attained in less than 300 milliseconds after ignition. This imposes a requirement on the igniter to instantaneously throw an ignition flame the entire length of the SRB. To complicate the system even more, there are two SRBs, spaced nearly 40 feet (12.4 meters) apart, that must be ignited simultaneously, and their output thrust must be closely matched to preclude the possibility of an uncontrollable trajectory of the space shuttle during launch. When fired, the pyrotechnics within each SRB S & A device must ignite the flame-generating pellets located directly below in a pellet basket. The pellets in turn ignite the initiative pyrogen which ignites the pyrogen igniter. The latter has the proper propellant burn characteristics to instantly project an initiating flame the entire length of the SRB. Several space shuttle SRB S & A device configurations were proposed, including that illustrated in Figure 7-1. In the following format, the encircled numbers identify a specific S & A device requirement. Following the number is a statement of the requirement, and, finally, a short explanation is given on how the fulfillment of that requirement was implemented into the design of the S & A device. Similarly encircled numbers in the figure direct the reader to the exact location(s) of the S & A device that fulfills that specific requirement.

① **Requirement:** The S & A device shall be designed to preclude hot gas leakage during and subsequent to SRB ignition.
**Implementation:** Thru-bulkhead-initiator (TBI) (see Chapter 3) cartridges preclude the necessity of rotating or sliding seals and isolate the end-firing detonator from both the igniter and the SRB internal operating environments.

② **Requirement:** The S & A device shall use two initiators that will function upon receipt of firing signals from two separate and independent circuits.
**Implementation:** Two end-firing detonators (see Chapter 3) are utilized with each being fired by separate and independent circuits. (Note: avonics associated with the S & A device have been omitted from the figures for clarity.)

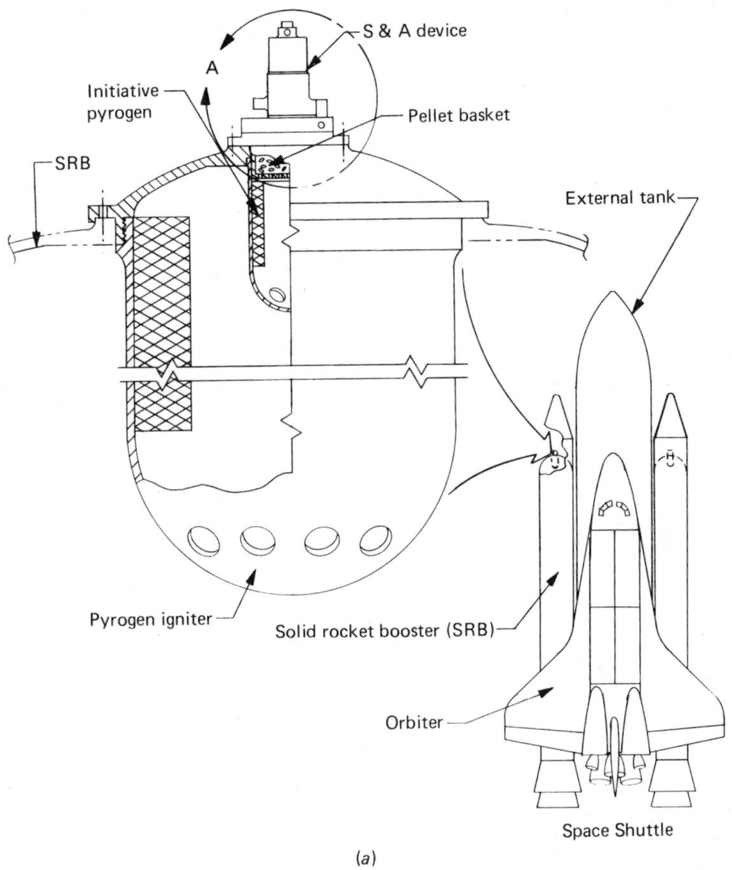


**FIG. 7-1.** Space shuttle SRB's ignition system. (*a*) Space shuttle configuration. (*b*) Cutaway of safe-and-arm device. (*c*) More S & A cutaways.

③ **Requirement:** The initiators shall be separable from the S & A device and if supplemental charges are required, all shall be separable from the S & A device.

**Implementation:** The end-firing detonators are removed from the rotor in the SAFE position; then when the rotor is moved to the ARM position the TBI cartridges are removable through the TBI cartridge ports in the rotor.

④ **Requirement:** The S & A device shall be able to enable and inhibit SRB ignition.

**Implementation:** In the ARM position, the end-firing detonators are aligned with the TBI cartridges located atop the pellet basket, thus enabling SRB ignition. Conversely, in the SAFE position, the end-firing detonators are out of alignment with the TBI cartridges, thus inhibiting SRB ignition.

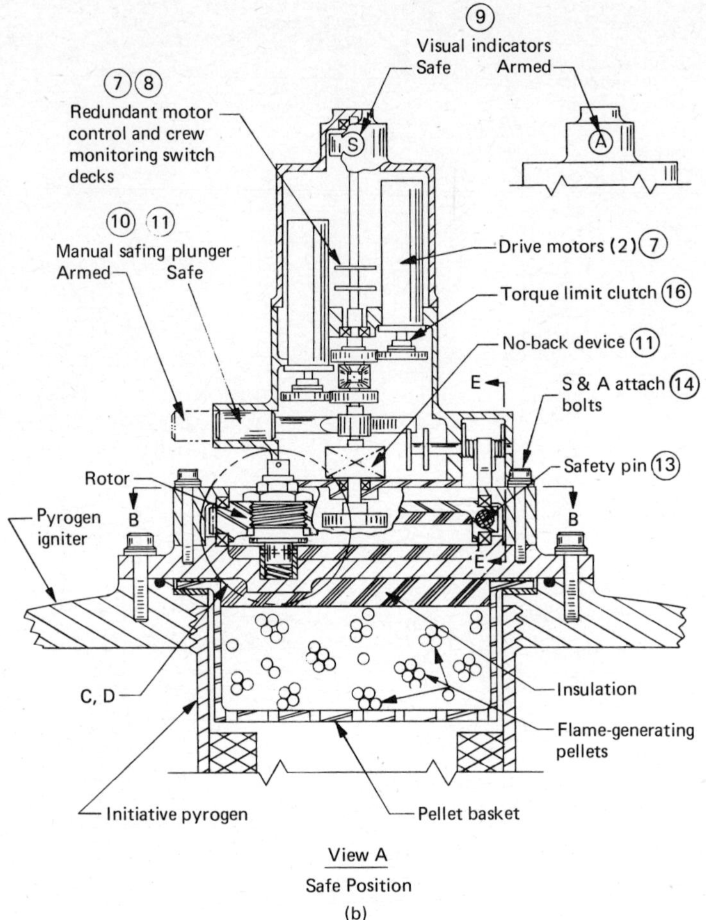


**FIG. 7-1.** Continued.

⑤ **Requirement:** The S & A device shall perform its intended function when either or both initiators are fired with the S & A device in the ARM position.
**Implementation:** In the ARM position, SRB ignition is enabled and each end-firing detonator is capable of initiating its respective TBI cartridge. Both TBI cartridges fire directly into a common pellet basket. The ignited pellet basket fires into the initiative pyrogen which fires into the pyrogen igniter that initiates the SRB.

⑥ **Requirement:** SRB ignition shall be inhibited when the S & A device is in the SAFE position and either or both initiators are fired.
**Implementation:** When the S & A device is in the SAFE position and either or both end-firing detonators are fired, the SRB will not be ignited because the detonatoɪs are out of alignment with their respective TBI cartridges.

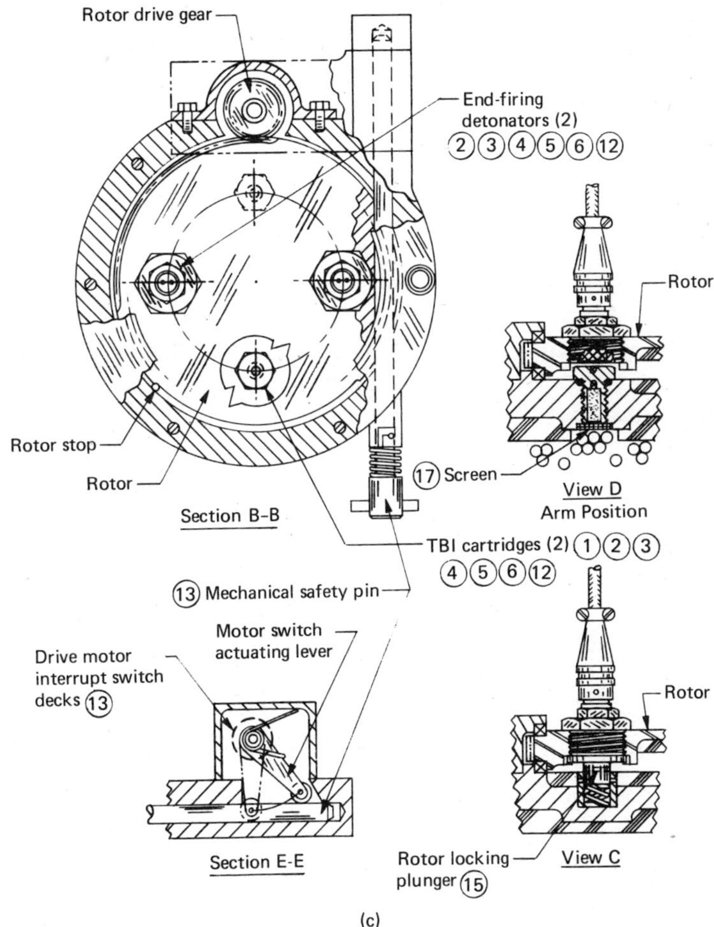


**FIG. 7-1.** Continued.

(7) **Requirement:** The S & A device shall provide positioning from SAFE-TO-ARM and ARM-TO-SAFE through redundant command, control, and actuation devices.
**Implementation:** Redundant command, control, and actuation motors to either ARM or SAFE the S & A device are provided.

(8) **Requirement:** The S & A device shall provide redundant remote position indicators.
**Implementation:** Redundant remote position indicator switches for both ARM and SAFE positions are provided.

(9) **Requirement:** The S & A device shall provide direct visual position indication in both the SAFE and ARM positions.
**Implementation:** Visual SAFE and ARM position indicators are provided.

(10) **Requirement:** The S & A device shall possess the capability for manual safing.
**Implementation:** Manual safing provisions are provided.

(11) **Requirement:** The S & A device shall be incapable of being manually armed.
**Implementation:** Manual arming provisions have been omitted, and a "no-back" mechanism permits only the actuators to drive the rotor to the ARM position.

(12) **Requirement:** The S & A device shall provide a 90-degree (minimum) out-of-line barrier without pyrotechnics in the barrier. (This requirement is peculiar to a particular type of S & A device that utilizes a movable barrier between pairs of fixed initiator and igniter cartridges.)
**Implementation:** This S & A device utilizes its housing as the barrier and rotates the end-firing detonators out of alignment with the TBI cartridges.

(13) **Requirement:** The S & A device shall provide a mechanical safety pin which prevents rotation of the barrier and mechanically interrupts the electrical power to the actuators when it is installed and which can only be installed in the SAFE position.
**Implementation:** Safety pin can only be installed when rotor is in SAFE position. Pin also actuates switches that interrupt power to the rotor drive motors.

(14) **Requirement:** The S & A device shall be separable from the igniter.
**Implementation:** The S & A device can be unbolted from the igniter.

Other features of the S & A device not specifically required are:

(15) S & A device cannot be subsequently armed if either or both end-firing detonators prematurely fire in the SAFE position.

(16) S & A device actuation motors are protected from overloading in event of overrun.

(17) S & A device provides screens between TBI cartridge ports and pellet basket to prevent foreign particles from entering TBI cartridge ports.

The author has taken some liberties in stating the aforementioned S & A device requirements. These adjustments merely were to aid the reader in comprehending the requirements and in no way altered their implementation into the S & A device presented.

## PYROTECHNIC PHOTOFLASH BULB

A recent introduction into the field of snapshot photography has been the nonelectric, pyrotechnic photoflash bulb that is marketed by several companies under the name of Magicube. Each Magicube contains four flashbulbs arranged in a square pattern on a camera mounting base. When the Magicube is installed on an appropriate camera, for example, any 110 pocket or "X" type (as shown in Figure 7-2*a*), one of the flashbulbs is positioned at the front of the camera. Around each of the four flashbulbs are individual reflectors that direct the flash toward the subject in front of the camera. A thin, transparent cover encloses the four reflectors and is bonded to the square perimeter outside the Magicube's circular mounting base. The lower surface of the Magicube base is provided with a center post that mounts with a camera to provide a means whereby the Magicube may be rotated 90 degrees after a bulb has been flashed so that another bulb may be advanced to the flashing position. Also, this action normally cocks the camera's shutter mechanism for the next picture. After four flash pictures have been taken with a Magicube, it must be removed and a new one installed before the next series of four photoflash pictures can be taken.

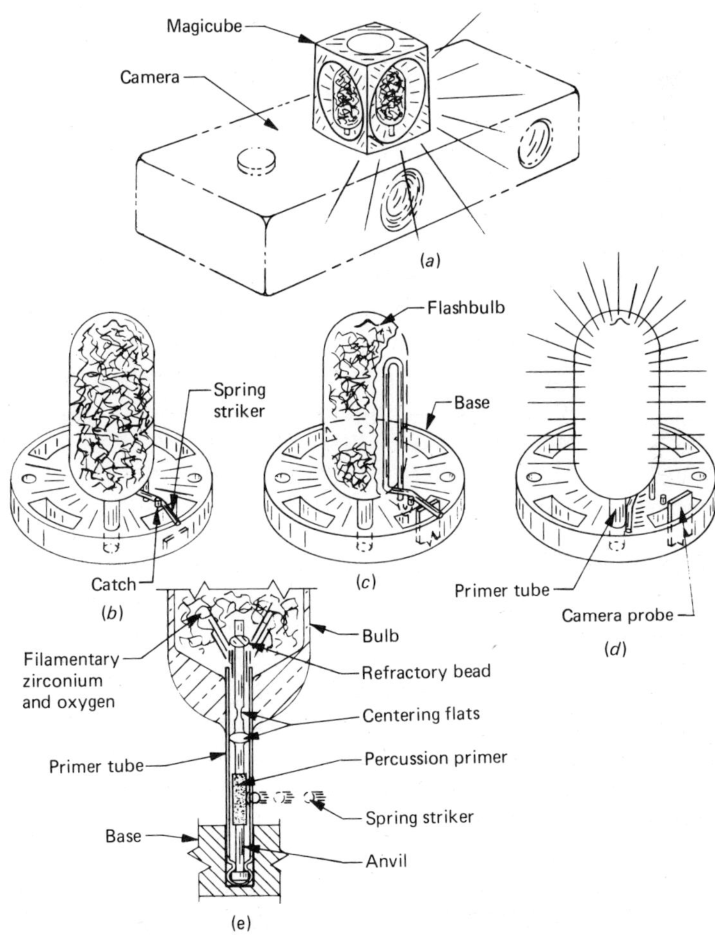


**FIG. 7-2.** Pyrotechnic photoflash bulb.

Figures 7-2*b* through 7–2*d* illustrate the sequence of events that synchronize the camera's shutter mechanism with the firing of the flashbulb. Only the center hub of the Magicube mounting base and one of four flashbulbs with its trigger mechanism are shown for clarity. To cause the bulb to flash, all that is required is to impact the base of the flashbulb with a small spring striker. This occurs as follows: The flashbulb is mounted in the circular hub of the Magicube base by a primer tube extending from its lower surface. A torsion spring striker, typically formed from 0.021-inch (0.53 millimeter) diameter music wire about 2.5 inches (63.5 millimeters) long, is shaped on one end to provide a catch for the cocked striker on the opposite end when mounted in the base. The cocked striker extends over an elongated hole that is also in the base of the Magicube. Displacement of the striker from the catch is effected by a blade-type probe extending from the camera through the elongated hole. Probe extension is synchronized with the camera's shutter mechanism. When the striker clears the catch, its torsion energy causes it to rapidly rotate and impact the primer tube of the flashbulb with sufficient kinetic energy to initiate the pyrotechnics within—causing the bulb to flash.

The internal configuration of the flashbulb is depicted in Figure 7-2*e*. The outer member is called the primer tube and interfaces with the glass bulb on the upper end and the Magicube mounting base on the lower. Centered inside the primer tube is an anvil consisting of a thin, stiff wire. The anvil has an upset head on its lower end around which is swaged the primer tube. The opposite end of the anvil is centered in the primer tube by two 90-degree adjacent coined surfaces. The anvil extends beyond the opened end of the primer tube inside the glass bulb. Near the anvil's free end is fused a glass refractor bead. The flashing material is a combustible, such as filamentary zirconium, in a combustion supporting gas, such as oxygen. Approximately midway between the anvil's upset head and coined protuberances, which centers it within the primer tube, is a band of percussion-sensitive pyrotechnic fulminating primer material. The fulminating primer material may consist of a single layer or multiple layers, but that used in percussive-type photoflash bulbs basically comprises a mixture of percussively reactive materials, a fuel and an oxidizer, arranged to ignite upon interparticle movement, such as a blend of red phosphorus and potassium chlorate. Intermixed with the percussively reactive material is a powdered, combustible, incandescible metal, such as titanium, which supplies the main zirconium ignition source. The flashbulb is mechanically fired by the impact of the spring striker against the primer tube in the area of the anvil's primer material. The striker indents the thin-walled primer tube so that the resulting impact pinches the primer material between the primer tube and the anvil, causing it to deflagrate upward through the primer tube, around the two coined anvil protuberances; it then impacts the refractory bead that spreads the flame and thoroughly ignites the zirconium, which results in the bright flash of light that is suitable for taking snapshot pictures.

## CREW ESCAPE SYSTEMS

Crew escape systems from aircraft and spacecraft utilize more, and a greater variety of, pyrotechnic devices than any other pyrotechnic system. Most military fighter and attack aircraft have crew escape systems permanently installed, whereas manned spacecraft and developmental aircraft have escape systems installed only during the flight testing phase of their development. When the aircraft or spacecraft becomes operational, the escape systems are removed except for the portion required for ground rescue capability.

The total capability of an escape system is determined by the analysis of a 6-degree-of-freedom computer program. The time-proven test of the accuracy of the computer program is to have it determine the results of a hypothetical set of escape parameters and compare them with the actual results of the same test performed on a test vehicle, which duplicates the crew escape portion of the actual aircraft or spacecraft that is accelerated by solid propellant rocket motors along a railed track. Such a sled test is shown in Figure 7-3 of the space shuttle orbiter's crew escape system. The escape system will accommodate two crew members: a commander, seated on the left side of the orbiter, and a pilot, seated on the right side. The system consists of severing 2 two-panel hatches and jettisoning them from the orbiter to provide holes through which two ejection seats are catapulted up dual pairs of guidance rails and out through the severed openings. The sequence of events, shown in Figure 7-3, is as follows: Part *(a)* shows a closeup of the test vehicle that duplicates the orbiter's flight deck mounted on a sled being accelerated along a track by another sled upon which are mounted several solid rocket motors (SRMs). The rockets are ignited at various stations along the track to produce the total velocity profile for this particular test. *(b)* The two flashbulbs on the simulated windshields indicate the instant the escape system is initiated at a velocity of 525 miles (845 kilometers) per hour. The flash on top of the orbiter is the outer jettisonable hatch panel's MDC severance system (see Chapter 4) cutting the ejection seat opening in the outer fuselage structure. An opening is simultaneously being cut in the inner crew module panel by an expanding tube assembly (XTA) severance system (see Chapter 4). Since the XTA is a confined severance system, no external evidence of its functioning is visible. *(c)* The two jettisoned hatches are visible at the left while the first crew member (the pilot) is seen clearing the top of the orbiter. The fire beneath the seat is the plume from the solid propellant rocket motor mounted on the back of the ejection seat. *(d)* With the jettisoned hatches still traveling nearly as fast as the sled, the drogue stabilizing parachute that prevents the ejected seat from tumbling can be seen not fully deployed just to the left of the pole-mounted box in the background. *(e)* The pilot's ejection seat drogue chute is fully deployed, its rocket motor has burned out, and the crew member/seat combination is coasting on a ballistic trajectory until: *(f)* A time-delay expires, severing the crew member/seat

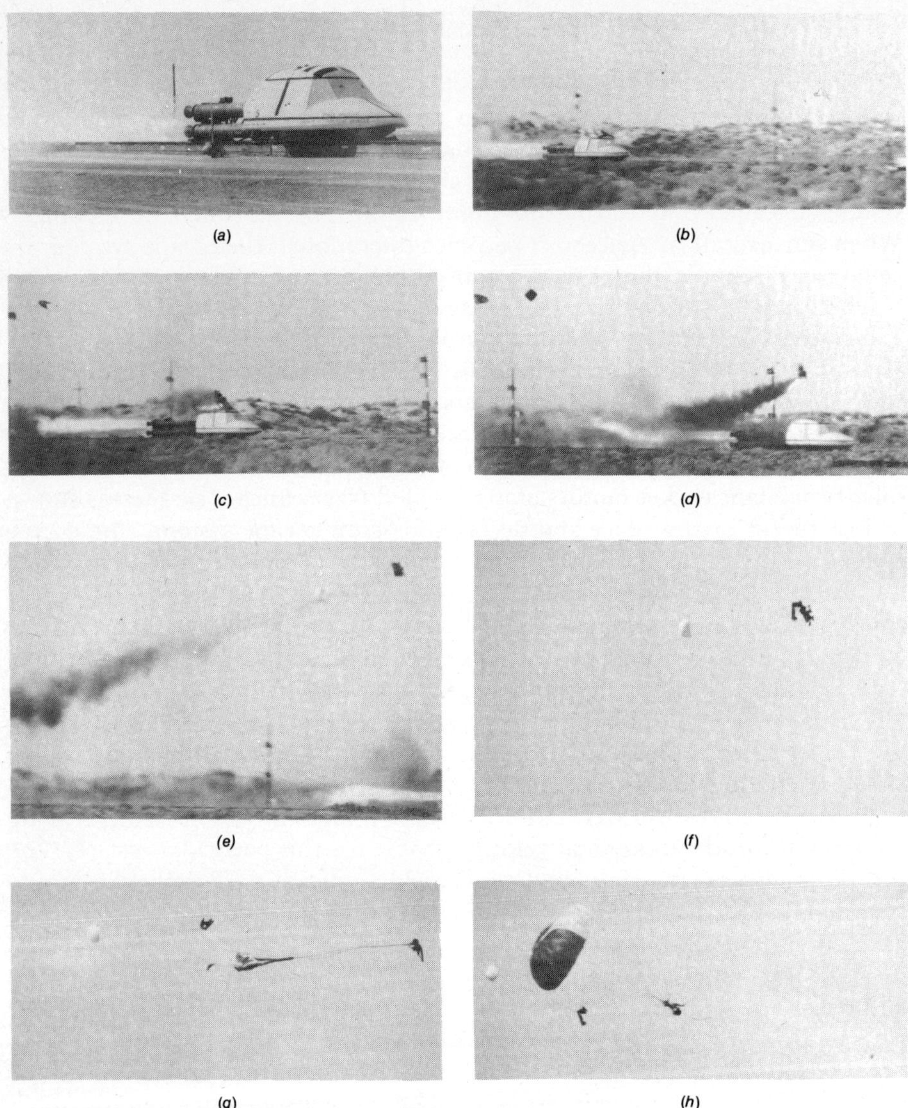

(a) (b) (c) (d) (e) (f) (g) (h)

**FIG. 7-3.** Space shuttle orbiter's crew escape system dynamic sled test. (*a*) Rocket sled ignition. (*b*) Hatches jettisoned. (*c*) Pilot's seat ejection. (*d*) Ejection seat clears orbiter. (*e*) Drogue chute stabilizes seat. (*f*) Pilot separates from seat. (*g*) Pilot's chute deploys. (*h*) Pilot's chute fully deployed.

strap retention system. The taut "butt-slapper" strap that ejects the crew member from the seat can be seen in the picture. *(g)* The pilot's recovery parachute can be seen being deployed with the empty ejection seat and drogue chute still in the background. *(h)* The final picture shows the pilot's recovery parachute fully deployed for the final lowering to the ground.

Sequence *(c)* through *(h)* is repeated, but not shown, for the orbiter's other crew member (the commander) 1/2 second later. The total elapsed time from escape system initiation until the commander is safely landed is slightly less than 15 seconds.

In appearance, a schematic diagram of a crew escape pyrotechnic system is similar to an electrical schematic. Figure 7-4 is the pyrotechnic schematic of the orbiter's crew escape system shown functioning in Figure 7-3. When the orbiter is on the ground, an emergency may arise from a crash landing that could jam the mechanically operated side entry hatch the crew would normally enter and exit through. In this case the crew has the option of jettisoning only the overhead hatches by pulling the hatch jettison T-handle initiator mounted on a console between the two crewmen. The hatch jettison system can also be initiated by a rescue team member by pulling a similar T-handle initiator mounted inside a door on the exterior of the fuselage. On military aircraft, the location of this handle is identified by a yellow arrow bordered in black with the word "RESCUE" printed on the yellow field in black letters. The stimulus from either T-handle initiator is transferred to

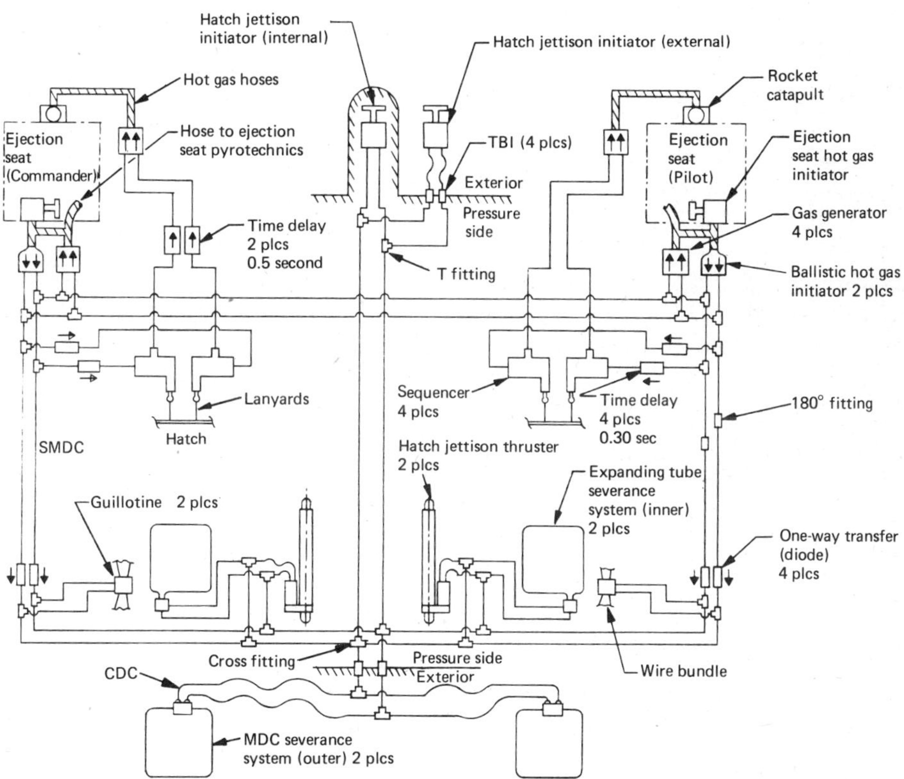


**FIG. 7-4.** Pyrotechnic schematic, space shuttle orbiter's crew escape system.

the severance and jettison systems of the 2 two-panel hatches by a redundant energy transfer system (ETS) consisting of shielded mild detonating cord (SMDC) and confined detonating cord (CDC) (see Chapter 4). The CDC from the externally mounted T-handle initiator must penetrate the pressure bulkhead of the crew module. This is accomplished by an O-ring-sealed thru-bulkhead initiator (TBI) (see Chapter 3), the donor being the external CDC and the acceptor being the internal SMDC. Once inside the crew module, the internal and external initiation stimuli interface at redundant ETS T-fittings. From this junction the stimulus is transferred along the same ETS to a pair of cross fittings. From there each ETS separates and journeys to the pilot's and commander's internal hatch jettison functions, where it initiates two crew module XTA panel severance systems, two wire bundle umbilical guillotines, and two hatch jettison thrusters. From the cross fitting the ETS interfaces with two more TBIs. On the exterior of the crew module each redundant ETS branches to initiate the two outer hatch panel MDC severance systems. Near the T-fittings where the ETS branches to the umbilical guillotines there are two one-way transfer (diode) devices in each ETS. These devices prohibit the ETS stimuli from extending on to the ejection seats and firing the rocket catapults.

Near the center of the pyrotechnic schematic are lanyards attached to redundant sequencers and to the jettisonable hatches. The purpose of these lanyards are to arm and fire the sequencers as the hatches are jettisoned from the orbiter. Note there is a second input to each of the sequencers coming from the ejection seats through redundant pairs of 0.30-second time delays. The third ETS port on each of the redundant pairs of sequencers is the initiating stimulus that fires the rocket catapults attached to each of the ejection seats. Two seat-ejecting conditions will be reviewed to illustrate why two stimuli inputs to the sequencers are required before the rocket catapults can be fired.

In the first condition the orbiter has just landed and stopped on the runway after returning from orbit. An emergency situation has arisen necessitating immediate crew ejection. Either crew member may pull the ejection seat hot gas initiator T-handle (see Figure 7-4). Unlike the ETS T-handle initiators previously discussed, these ejection seat T-handles initiate redundant gas-generating cartridges that perform numerous functions on the ejection seats prior to initiating the rocket catapult that ejects the seat and crew member from the orbiter, such as retracting the crew member's seat back-pad, which allows the crew member to be moved farther back in the ejection seat, retracting the crew member's shoulder harness, and retracting the crew member's feet from the rudder control pedals back into the ejection seat envelope. At the same time the hot gas is initiating the ETSs where they interface at the ballistic hot gas initiators. Each ETS redundant output branches to the other crew member's ejection seat and transfers back into hot gas at the gas generators. Simultaneously, the ETS continue to a second branch that interfaces with the previously mentioned 0.30-second time delay. The other branch continues through the one-way transfer (diode) and on to the hatch jettison systems, both previously discussed. Since each of the jettisonable hatches weighs over 300 pounds (136 kilograms), the hatch jettison thrusters require over 200 milliseconds to jettison the hatches. If the 0.30-second time delays were not in the system, the stimuli from the ballistic hot gas initiators would pass through the sequencers and the second pair of gas generators and fire the rocket catapults before the hatches had sufficient time to be jettisoned clear of the seat-ejection paths. The addition of the 0.30-second time delays provides sufficient time for the hatches to be jettisoned clear of the seat-ejection paths. It is this condition of the orbiter with no forward velocity that requires the longest time for the hatches to be jettisoned.

The second condition is with the orbiter having a forward velocity in excess of 75 miles (121 kilometers) per hour and below an altitude of 75,000 feet (22,860 meters) at the time of the emergency necessitating crew ejection. This time the hatch jettison thrusters do not eject the hatches because when the forward edges of the outer fuselage panels are severed and exposed to the extremely high dynamic air loads, the panels are immediately rotated about their hinges on their aft transverse edges at a rate many times faster than that normally provided by the hatch jettison thrusters. If the 0.30-second time delays were not provided, the fast jettisoning hatches would fire through the sequencers and fire the rocket catapults before the preejection seat functions had time to be performed. In summary, it takes both the event of the hatches being jettisoned and the time of 0.30 second to occur in either sequence before the rocket catapults can be safely initiated to eject the crew member from the orbiter. Since both the commander's and pilot's ejection seat rails are mounted parallel, there is the possibility of the two ejecting seats colliding due to dissimilar weight and aerodynamic characteristics of the two ejecting masses. To preclude this possibility, redundant 0.5-second time delays are installed in the ETS lines between the sequencers and the rocket catapult of the commander's ejection system.

Figures 7-5*a* and 7-5*b* are views of the orbiter's pilot and commander stations respectively, looking aft and down through the hatch openings of the escape system test vehicle at the conclusion of the test run sequenced in Figure 7-3. In the photograph an extra ejection seat and dummy pilot had been temporarily placed in the pilot's preejection position for reference purposes only. The severed edges of the inner and outer hatch panels, as well as the redundant CDCs that initiate the outer panels' MDCs severance systems, are clearly visible in both Figure 7-4 and Figure 7-5. The CDCs are mounted between the orbiter's outer structure and the inner crew module. Hatch jettison thrusters are mounted between the ejection seat's guide rails on the rail's support structures. On the outboard sides of each rail support structure are the redundant sequencers. In the center of the commander's station, Figure 7-5*b*, is the barrel portion of the rocket catapult. Considerable smoke and rocket propellant residue have been deposited inside the test vehicle from the ignition of the ejection seats' rockets (see Figures 7-3*c* and 7-3*d*).

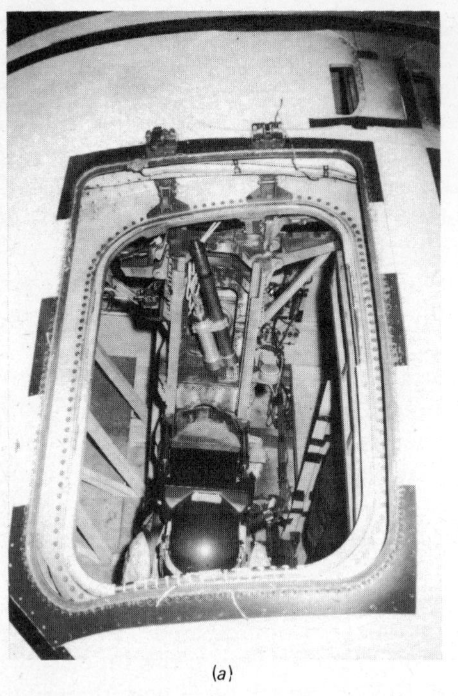

(a)

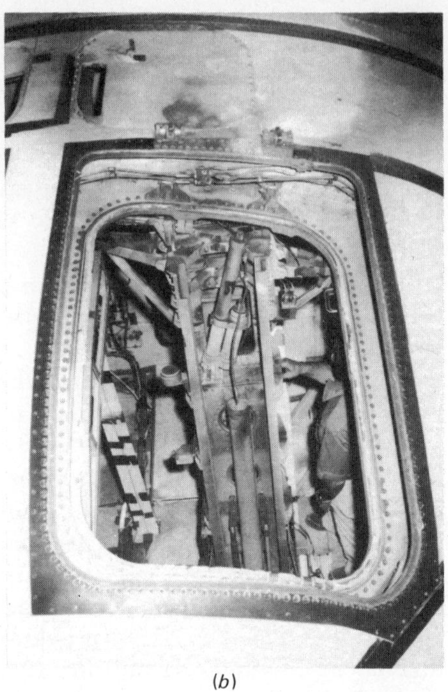

(b)

**FIG. 7-5.** Orbiter's severed hatch openings. (*a*) Pilot's hatch opening. (*b*) Commander's hatch opening.

Figures 7-6*a* and 7-6*b* are pretest, aft end views of the rail support structure for the commander's and pilot's ejection seats respectively. On the upper outboard sides of both structures are mounted the guillotines with dummy umbilicals threaded through them. Just below the guillotines the sequencers are mounted. Even the lanyards from the sequencer to the hatch have been installed on the commander's side, whereas they have not yet been installed on the pilot's side. Also on the commander's side, the pair of horizontally mounted, silver-colored, tubular objects with ETS lines protruding from either end are the 0.30-second time delays, whereas the vertically mounted similar pair are the 0.5-second time delays. SMDCs, CDCs, 180-degree fittings, cross fittings, T-fittings, etc., are evident in both photographs. Although these exposed ETS lines appear vulnerable, closeout covers are installed over the entire ETS. Two such covers can be seen on the lower back surface of the pilot's rail support structure and one on the commander's.

## FRICTION INITIATED DEVICES

In addition to the many initiators described in Chapter 3, there is a group of devices that are initiated by merely rubbing them briskly enough to raise the temperature of their heat-sensitive priming material to its kindling point.

(*a*)

(*b*)

**FIG. 7-6.** Orbiter's ejection seats rail support structures and energy transfer system. (*a*) Commander's station. (*b*) Pilot's station.

The invention of the match in the late nineteenth century coincided with the need by the developing railroads for a signaling and warning fusee (flare) to control the safe operation of numerous trains on common tracks.

The similarities of matches with fusees are many. They differ primarily in their relative sizes and final burning agents, for example, wood or cardboard for matches and red-flame producing propellant for fusees.

## Matches

People in all walks of life are familiar with one or two specific pyrotechnic devices, for example, fireworks, dynamite, and ammunition, but the one group of pyrotechnics enjoying the broadest popularity is by far—matches. Two types of matches have evolved: the *strike anywhere* match and the *safety* match. The evolution of these two types of matches is a story of painful suffering and generosity that merits review if the reader is to fully appreciate the relatively simple, commonplace match.

Primitive peoples started their fires with a drill consisting of an upright stick that was rapidly rotated in a socket of dry wood. Later, people learned to strike a piece of flintstone with another stone called pyrite to create a spark sufficiently hot to cause dry leaves and grass or wood shavings to ignite. A slight improvement was realized when the pyrite was replaced with a piece of steel bar. Fire making was so troublesome that a common household utensil was a long-handled metal box with holes in the lid that was used to transport hot coals from one place to another. Fire starting was extremely difficult and time-consuming, and it required much less effort to maintain a fire continuously than to start a new one.

In 1680, the English chemist Robert Boyle manufactured and sold a packet consisting of coarse sheets of paper coated with phosphorus and splinters of wood with one end tipped with sulfur. When the sulfur-tipped wood splinter was rapidly drawn between a fold of the phosphorus-coated paper, the sulfur tip would burst into flame. A sufficient amount of sulfur was provided to sustain the flame long enough to ignite the wood splint. Since the price of phosphorus in those days was extremely high, only the wealthy could afford these early matches, and in no time even their interest in this novelty waned. Soon after, both sales and experimentation in match development ceased.

It was nearly a hundred years later, in France, that the phosphoric candle or ethereal match appeared. It consisted of a strip of heavy paper, about 2.25 inches (62.4 millimeters) long, with one end dipped in phosphorus. The entire match was then encased in a thin-walled tube of glass, sealed at both ends, which resembled a present-day oral thermometer. To ignite the match the user would simply grasp the glass match between thumb and forefinger of both hands then squeeze and bend it until the glass broke. Then the two pieces of glass were pulled apart, exposing the phosphorus-tipped end of the match to air, causing it to ignite and also set the exposed paper afire.

The Englishman John Walker is credited with inventing the early versions of our modern-day friction matches. They were of the strike anywhere type made of wood splints tipped with a compound composed of potassium chlorate, antimony sulfide, and gum arabic (the sap of the Acacia tree). They were unreliable and difficult to strike. To counter these shortcomings, France's Dr. Charles Sauria substituted white phosphorus for the antimony sulfide of Walker's formula. The white phosphorus set off a wave of necrosis that was to exact a fearful toll of suffering and lives, not only of match makers but users as well, until past the end of the nineteenth century. The necrosis attacked the bones. Particularly subject to damage were the bones of the jaw, which were attacked through cavities in the teeth, and the condition was soon referred to as "phossy jaw" by workers. Its results were severe disfiguring, usually followed by death. In addition, innocent persons, especially babies, were poisoned by ingesting phosphorus match heads. The heads also provided an easy source of poison for suicide and even murder.

"Phossy jaw" plagued match factories and the public until 1900, when the Diamond Match Company bought the French Sevenne-Cahen patent for sesquisulfide of phosphorus, a nonpoisonous substitute for white phosphorus. The Diamond Match Company was soon to discover that sesquisulfide of phosphorus would not work in American atmospheric conditions. In 1911, William A. Fairburn, a mechanical and chemical engineer who became president of the Diamond Match Company in 1915, discovered that adding chlorate of potash to the formula adapted it to American climatic conditions. The nonpoisonous formula was presented to the United States government for use by all rival match companies, a humanitarian gesture that won public commendation from President William Howard Taft and for which Mr. Fairburn and his company were given the Louis Livingston Seman Medal "for the elimination of occupational disease."

The new formula raised the ignition temperature of the tip compound to 340°F (171°C), a considerable increase in safety factor. It also ended fires caused by rodents. Experiments conducted by Mr. Fairburn proved that while rats and mice would gnaw on phosphorus matches, sometimes igniting them, they would not touch the new match heads even if near starvation. He also invented a spraying device that soaked splints with a chemical solution, ammonium phosphate, to deaden the "afterglow" of burned matches. This too went into the public domain. Other Fairburn improvements were an odorless compound for sulfur and a new method of seasoning Idaho white pine for splints. Improvements by others included soaking the ends of matchsticks which carry the heads in hot paraffin wax before the head is put on. This allows the flame to be transferred from the match head to the stick. Without the parrafin wax the match would go out as soon as the combustible head was burnt.

Book or safety matches were invented in 1892 by Joshua Pusey, a Philadelphia patent attorney. The heads of these matches contain chlorate of potash and sulfur and will only ignite when struck on red phosphorus or are subjected to a temperature sufficient to ignite them. Pusey's books contained

50 matches and had their red phosphorus striking surface on the inside where sparks frequently ignited the remaining matches, a danger quickly corrected by the Diamond Match Company after it bought Pusey's patent in 1895. Unpopular at first and made entirely by hand, book matches became big business in 1896 when a brewing company offered 10,000,000 books to advertise their brews, forcing creation of machinery for swift production in volume. Figure 7-7 presents the modern versions of the three basic types of matches. Note that safety matches are of three different kinds—book, and small- and large-size wooden matches in boxes.

Since the beginning of the twentieth century, most improvements in the match industry were made in manufacturing methods. Not until World War II did the next major match advance occur. The United States War Department needed a match that would function in the long rainy seasons of the South Pacific and called upon the industry for help. Raymond D. Cady, chief chemist for the Diamond Match Company, developed a match that can be submerged in water for up to 8 hours and still function. The match was supplied at a rate of more than 10,000,000 per day to the armed forces.

The United States makes 300 billion each of paper book and wooden "strike anywhere" matches and 100 billion wooden safety matches annually, the largest national output in the world. Other large manufacturers are Great Britain, Russia, and Sweden. During recent years there has been a gradual reduction in the production and use of wooden matches. Rural electrification wiped out a major market, while the increase in cigarette

**FIG. 7-7.** Modern strike anywhere and safety matches.

consumption brought a compensating demand for book matches. Only about 5 billion matches are imported each year into the United States, primarily from Sweden, Italy, Finland, and Japan.

**Fusees (Flares)**

In the late nineteenth century, a need arose for a warning and signaling device in the nation's developing railroads. Known as *fusees* in the railroad industry and as *flares* by the motoring public, these signaling devices consist basically of a waterproof cardboard tube having a crimped closure at the base that is filled with a red-flame-producing pyrotechnic propellant consisting typically of approximately 74 percent strontium nitrate, 6 percent potassium perchlorate, 10 percent sulfur, 4 percent grease or wax, and 6 percent sawdust and/or hardwood shavings. The fusee's ignition system is similar to that of the safety match: the rubbing surface of a removable cap is coated with red phosphorus, which is rubbed on a head of chlorate of potash and sulfur that is in intimate contact with the red-flame-producing propellant.

There are five basic types of fusee, with 5-, 10-, 15-, 20-, and 30-minute burning times. Figure 7-8 shows three of these types. The 5- and 10-minute varieties are used primarily with railroads, the 15-minute type is a flare for general use by motorists, the 20-minute fusee is used by over-the-road truckers, and the 30-minute type is for police departments. Fusees are approximately 1 inch (25.4 millimeters) in diameter, and their length is approximately 0.5 inch (12.7 millimeters) per minute of burning time plus

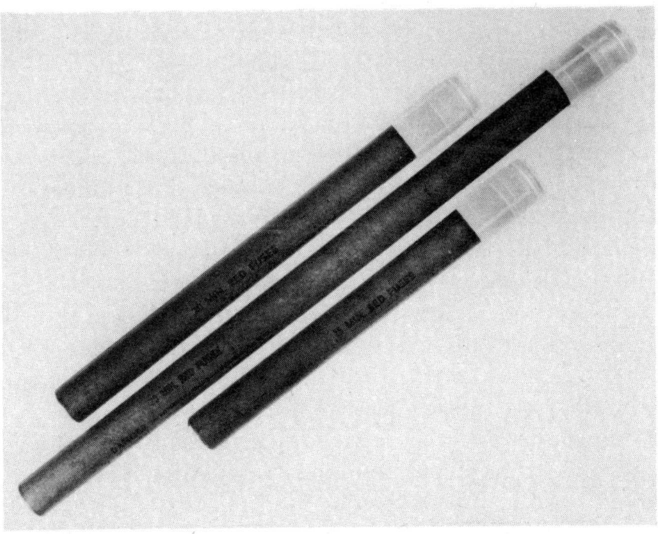

**FIG. 7-8.** Signal and warning fusees.

approximately 1.5 inches (38 millimeters) for end closures and ignition compounds. Railroad fusees also have pointed spikes built into their bases. This allows them to be thrusted into the wooden rail ties, thereby providing greater illuminated visibility in an upright position. Road flares also are available with wire bipods to support the ignited flare in a similar upright position on a hard roadway surface.

Fusee performance is controlled by the Association of American Railroad's Bureau of Explosives. In addition to specifying precautionary labeling, the bureau stipulates the physical strength of fusees and requires a waterproofing performance demonstration in which the fusee is immersed in water for 10 minutes and then immediately ignited by the normal means of striking the cap across the ignition head. The minimum burning time of the fusee is that stated on the labeling, with the maximum time being approximately 3 minutes more. Light intensity shall not be less than 70 candlepower when measured with a photometer not less than 24 inches (61 centimeters) from the fusee. Another test requires that a fusee be ignited in the usual manner and then submerged in water in a vertical position, head down, for 1 minute without the flame being extinguished.

Safety flares are carried in the trunks of millions of cars and cabs of commercial trucks as a highway warning device that is convenient to carry and is instantly available for emergency use on highways in any weather condition—wind, rain, or snow. Their use reduces the possibility of collisions caused by vehicles parked or stalled on the highways. They provide ample light for changing tires and are invaluable in emergency situations. They are recommended for use by the National Safety Council, state highway patrols, and county and municipal police departments.

## HOT PATCHES

In the twentieth century all types of mechanized modes of transportation were ushered in, including the automobile, airplane, motorcycle, and bicycle, that soon necessitated ways of providing a smoother, more comfortable ride. Along with improvements in suspension systems, the dominant contributor was the pneumatic tire. However, with this advancement also came the problem of occasional punctures from road hazards such as chuck holes, nails, glass, sharp stones, etc. Nails and other small, sharp objects are the greatest cause of tire punctures. The immediate resolution of a "flat tire" is, of course, to replace the tire with the spare in the trunk. Sooner or later the flat tire will have to be repaired or replaced if even more inconvenience is to be avoided should a subsequent flat tire occur.

Although numerous methods of tire repair are available, the "hot patch" is the most durable and reliable method for small punctures. To illustrate the simplicity of installing a hot patch, Figure 7-9 presents the sequence of operations of repairing a bicycle inner tube that has been deliberately punctured with a nail. Figure 7-9*a* shows in the background the tube with

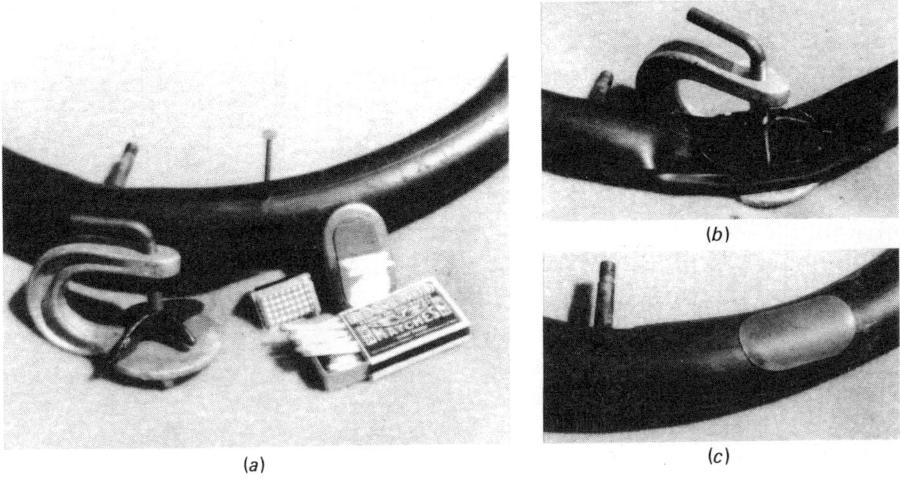


**FIG. 7-9.** Hot patch inner-tube repair. (*a*) Punctured tube and repair kit; (*b*) Vulcanizing process; (*c*) Puncture repair completed.

the nail still inserted and the necessary tools and materials for repairing the puncture in the foreground. They consist of a screw clamp to hold the patch in intimate contact with the tube being repaired, a buffing tool to roughen the area around the puncture for better patch adhesion, and a patch assembly consisting of a shallow sheet-metal cup filled with a flameless, heat-producing pyrotechnic fuelboard. On the bottom of the cup is a thin layer of uncured rubber approximately 0.04 inch (1 millimeter) thick, with the protective poly film cover partially removed for clarity. Matches or some other high-temperature heat source is required to ignite the pyrotechnic fuelboard at the proper time. Figure 7-9*b* shows the hot patch installed over the roughened puncture with the screw clamp holding it tight against the tube with the fuelboard, having been ignited, still burning. Heat from the combusting fuelboard heats the metal cup, thereby raising the temperature of the rubber sufficiently to vulcanize or cure it, i.e., at 350°F (177°C) for approximately 25 seconds. The clamp is retained for approximately 4 or 5 additional minutes to allow the latent heat of the cup to complete the vulcanizing process. Uncured rubber has the texture of putty, and compounding it with sulfur and lesser amounts of other chemicals, followed by the simultaneous application of pressure and heat, vulcanizes or cures the rubber into any desired consistency, depending upon its final application. Figure 7-9*c* shows the completed patch installation. The tube has been partially inflated to illustrate the hot patch is ready for immediate use.

The repair procedure for a large hole or small tear is to remove a small amount of uncured rubber from another patch and fill in the hole or tear before covering the puncture with the final repair patch.

Hot patches are available in round, oblong, and diamond shapes. They have also found widespread use in repairing rubber shoes and boots, gloves, aprons, hot water bottles, sport-ball bladders, tubeless tires, boats, and many other rubber articles.

## INFLATION SYSTEMS

There are three basic types of pyrotechnic initiated inflation systems: The first utilizes a compressed gas, usually nitrogen, stored under high pressure, inside a storage bottle or tank. At the outlet end of the tank is fitted an explosive valve (see Chapter 6) that, when initiated by a pyrotechnic cartridge, will open the valve, enabling the gas to flow into the device to be inflated. The second type employs a solid propellant that, when initiated, generates a large volume of relatively cold-temperature gas, mostly nitrogen, with sufficient pressure to inflate an air bag used to protect the occupants of an automobile during a crash. The third type is an augmented, cool-gas inflation system that combines the rapid, gas generating capability of a pyrotechnic propellant with an aspirator to pump ambient air into large-volume emergency escape slides of modern, wide-bodied jet airliners. The two latter inflation systems are explained in greater detail in the following discussions.

### Automobile Passive Restraint Systems

Since World War II, the United States Department of Transportation has mandated many now familiar standard automobile safety equipment features, including the collapsible steering wheel column, padded instrument panels and sun visors, energy-absorbing safety glass windshields, dual braking systems, and head restraints. These safeguards have prevented thousands of deaths and serious injuries to occupants in automobile crashes since their implementation.

On June 30, 1977, the Federal Motor Vehicle Safety Standard 208 was amended to eventually require every new automobile purchased in the United States to be equipped with passive crash protection for all front seat occupants. The amendment goes into effect September 1, 1981. A passive crash protection system must meet two basic requirements. First, it must protect all front seat occupants during frontal and front-angular impact forces equivalent to an automobile crashing into a fixed object at velocities up to 30 miles (48.3 kilometers) per hour. Second, the system must be completely passive, that is, provided by the automobile itself without need for any physical action by the occupants.

To date, automotive safety engineers have produced two systems that will fulfill the requirements of Standard 208—automatic seat belts and automatic air bags. The standard does not require or identify any specific approach with regard to the selection or development of the protection system. The choice is left entirely to the individual auto maker and the new car buyer.

The Transportation Secretary ordered a 3-year phase-in, requiring passive protection in all full-size 1982 model automobiles, in all intermediate-size

1983 models, and in all 1984 models. An estimated 12,000 lives will be saved and over 100,000 major injuries averted per year. Passenger vans, pickup trucks, and certain recreational vehicles are not presently included in the standard. There is a possibility that several auto makers will offer both automatic belt and air-bag systems as extra-cost options before the passive restraint system law goes into effect. A brief description of the automatic belt system is provided so the reader may compare it with the other leading contender—the automatic air bag. It should be noted that currently both automatic restraint systems afford protection only for the front seat occupants of an automobile. Department of Transportation statistics show that 92 percent of all those killed in crashes are in the front seats. In all but the most violent crashes, rear seat passengers receive protection from the front seats and, in some cases, from lap belts.

**The Automatic Belt Restraint System** A shoulder belt encloses the driver and the right front passenger as they close their respective doors. When either door is opened, its outward and forward swing carries the upper outboard end of the belt away from the back of the seat, allowing the occupant to slide with relative ease between the seat and the belt. As the door closes, a storage reel attached to the lower center area of the car takes up the slack in the belt, which lies diagonally across the chest of the occupant. A knee restraint prevents the occupant from sliding under the belt in a frontal crash. The belt is designed to allow reasonably free movement while driving or riding in the car. The reel mechanism is designed to detect either abrupt acceleration of the occupant or abrupt deceleration of the car, both of which occur in a frontal crash. The storage reel then locks in position, freezing the belt length.

The belt can be unfastened at the buckle on the door for quick exit after a crash or for any other reason. Some systems prevent restarting the car's engine once it has been shut off unless both belts are again fastened into place. Others provide a sound or light warning that a belt is disconnected. An active lap belt is being considered at least by one auto maker as a supplement to the passive belt for each front seat occupant.

Present technology has not yet developed an automatic belt system for the center position of the three-person bench seat.

**The Automatic Air-Bag Restraint System** The air-bag protection system is composed of a few basic parts (see Figure 7-10). When the car's ignition switch is turned on, the indicator light glows for 6 or 8 seconds. This lamp is the only component of the air-bag system visible inside the car. During this interval, a diagnostic circuit checks the operating condition of the system. At the same time, a capacitor is charged which initiates the system if the car's battery is destroyed during the initial impact of the crash. When the indicator lamp fades out, the capacitor is charged and the system is assumed to be fully operational.

When a frontal or front-angular crash generates impact forces equivalent

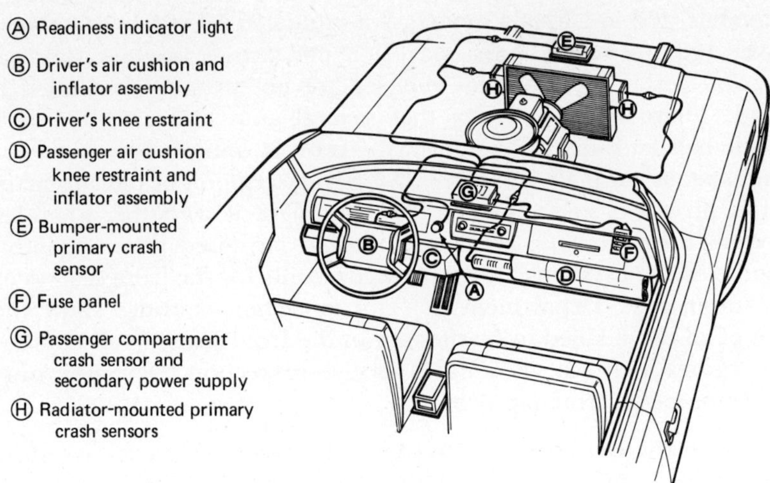


**FIG. 7-10.** Air-bag system schematic.

to hitting a fixed object at about 12 miles (19.3 kilometers) per hour, the abrupt deceleration causes the impact sensors to make electrical contact that initiates the air-bag system. Figure 7-11 shows the basic concepts of two such impact sensors. Figure 7-11*a* is a directional sensor that makes contact when the automobile in which it is mounted impacts a barrier while traveling from left to right. The sensor consists of two electrical contacts—one from the power source (the capacitor) and the other to the air-bag inflators that protrude inside a chamber at the opposite end of a weighted disk that is spring-loaded away from the contacts. When the sensor is impacted abruptly against a barrier with sufficient velocity, the momentum of the disk causes the retention spring to stretch a sufficient amount to allow the disk to impact the two contacts at the end of the chamber. The disk, being made of an electrical conductive material, completes the circuit across the contact and thus triggers (initiates) the air-bag system. As can be seen, this type of sensor is immune to all impacts except those nearly parallel to the axis of the sensor.

At the opposite extreme from the sensor with directional sensitivity is the omnidirectional impact sensor shown in Figure 7-11*b*. This type consists of a weighted pendulum made of stiff music wire whose longitudinal axis is mounted perpendicular to the horizontal axis of the car. When the sensor is impacted from any direction that is approximately perpendicular to its axis with enough force to cause the pendulum to bend sufficiently to contact the electrical conductor lining the inner wall, the sensor completes the circuit through the pendulum and thus triggers (initiates) the air-bag system. A pendulum impact sensor can be made directional to any sector or sectors desired by simply interrupting the contact liner where circuit completion is undesired.

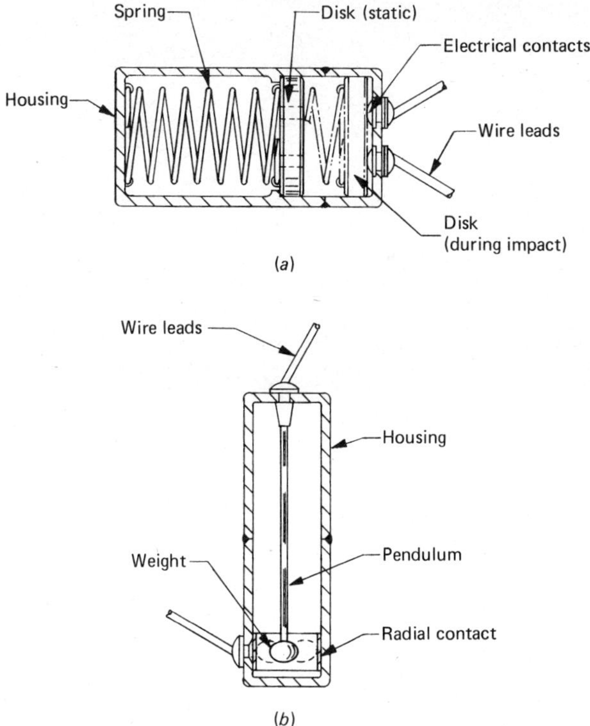


**FIG. 7-11.** Impact sensors. (*a*) Directional sensor. (*b*) Omnidirectional sensor.

Sensors are usually mounted in two locations on an automobile: near the front, on the bumper, grille, or radiator, and on the forward bulkhead (firewall) of the passenger compartment.

The knee restraints are designed to keep front seat occupants from sliding under the inflated air bags during frontal crashes. They hold the lower portion of the body in place so the upper portion remains properly positioned for air-bag cushioning.

Some air-bag system electronics include a "safing discriminator." This device is a deceleration detector designed to prevent inadvertent inflations, for example, slamming a door or light "parking lot" encounters, by requiring abrupt deceleration of the car before the impact sensors can send the initiating signal to the air bags.

The air bags themselves are the heart of the automatic air-bag restraint system. In the front seat are mounted two air bags, as shown in Figure 7-12*a*. One is concealed in the hub of the steering wheel to protect the driver and the other is concealed in the instrument panel to protect a maximum of two passengers. Upon sufficient impact, as previously mentioned, the abrupt deceleration causes the sensors to initiate the inflators built into each air bag.

(a)

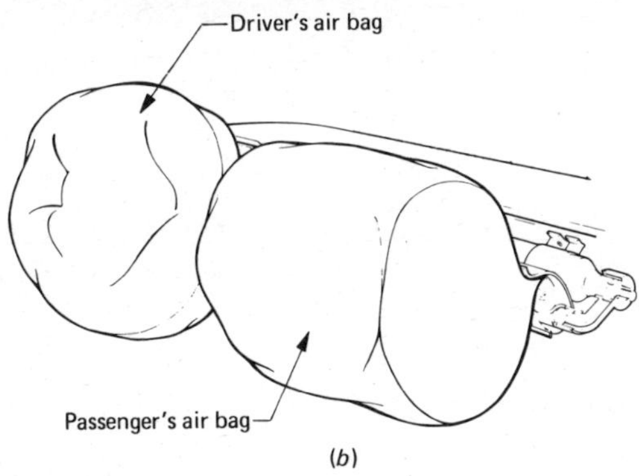


(b)

**FIG. 7-12.** Automatic air-bag restraint system. (*a*) Air bags installed in 1974 Oldsmobile. (*b*) Driver's and passenger's air bags fully inflated.

As they inflate, the bags break out from their concealment covers, filling instantly to form soft pillows between the occupants and the dashboard of the car, as shown in Figure 7-12*b*. The cushions are fully inflated and actually begin to deflate in about 1/25 second. That means the process occurs so rapidly that if you were to blink your eyes, you would miss the air-bag inflation sequence completely.

Figure 7-13 is an exterior photo at the instant an automobile, driven by professional stunt driver Vic Rivers, crashed into a concrete block wall. Note how the momentum of the car nearly lifts the rear wheels from the ground. Included in the instrumentation was an interior high-speed motion picture camera that recorded the total crash sequence. Selected frames of the three main events of this particular crash are presented in Figure 7-14: *(a)* initiation of the air-bag inflator; *(b)* rapid auto deceleration, causing the driver to lunge forward into the air bag against the steering wheel; and *(c)* the driver

**FIG. 7-13.** Air-bag-equipped car crashing into barrier.

rebounding from the air bag back into the seat. The main purpose of the rapid deflation of the air bag is to reduce the force of the rebound. Rivers suffered only minor abrasions caused by his violent deceleration into the air bag.

A detailed view of a driver's air-bag installation in a steering wheel is shown in Figure 7-15. The most apparent external difference is a slightly larger hub to accommodate stowage of the compacted air bag. The inflated volume of the air bag is approximately 2 cubic feet (57 liters). The open end of the air bag is attached to the hub of the steering wheel around an interfacing airtight seal. The air bag is folded back and forth (accordion style) over itself until it is as compact as it can be and still assure subsequent rapid inflation. To protect the air bag during everyday automobile use, a breakaway, protective shroud is folded over its exposed upper surface. Snap-open locks, molded into the shroud, retain it in a closed position until the initial inflation pressure within the air bag opens the locks, allowing the air bag to be ejected out of the steering wheel hub and thence inflated to its maximum pressure of approximately 5 pounds per square inch (0.35 kilograms per square centimeter), at which time the pressure begins to subside.

Also mounted to the steering wheel hub, inside the opened end of the air bag, is the air-bag inflator. The cross section of a typical inflator is shown in Figure 7-16. When the impact sensors (see Figure 7-11) undergo abrupt deceleration of sufficient magnitude to cause closure across the contacts, an electrical capacitor is discharged through the bridge wire of the ignitor cartridge whose flaming output is directed through holes in the cartridge

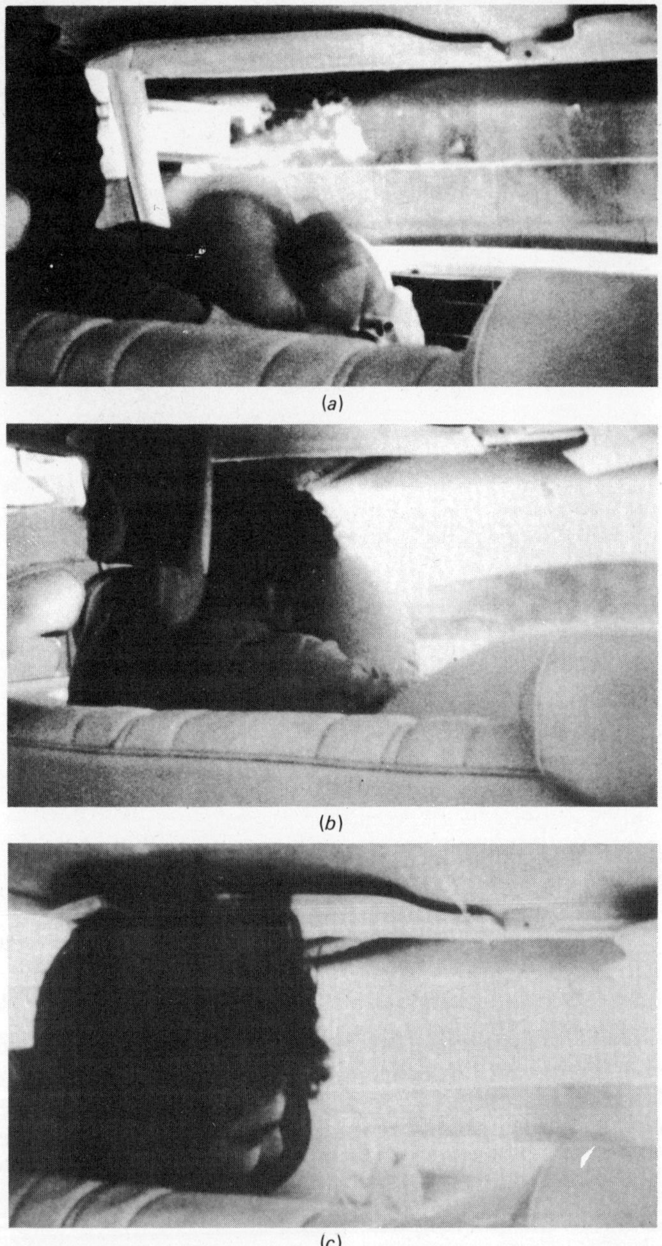

(a)

(b)

(c)

**FIG. 7-14.** Interior views of driver during crash of Figure 7-13. (*a*) Impact initiates inflator and air bag emerges from steering wheel hub. (*b*) Rapid deceleration causes driver to lunge forward into air bag. (*c*) Driver rebounds unharmed from air bag back into seat.

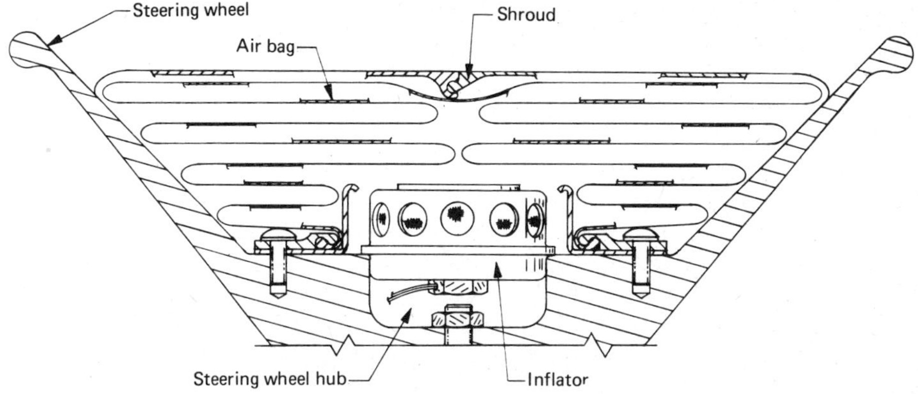


**FIG. 7-15.** Driver's air bag and inflator in steering wheel hub.

cup into the combustor filled with gas-generating pyrotechnic propellant "tablets." Gas formed by these deflagrating tablets traverses radially through multiple filters. The efficient filtration of the effluent gas is necessary because the pyrotechnic combustion of the tablets also produces a "fume" and ash of solid residue which must be removed from the gas by filtration. The filters used remove the particulate matter by condensation and adhesion and by mechanical trapping. They are also characterized by small pore size, high

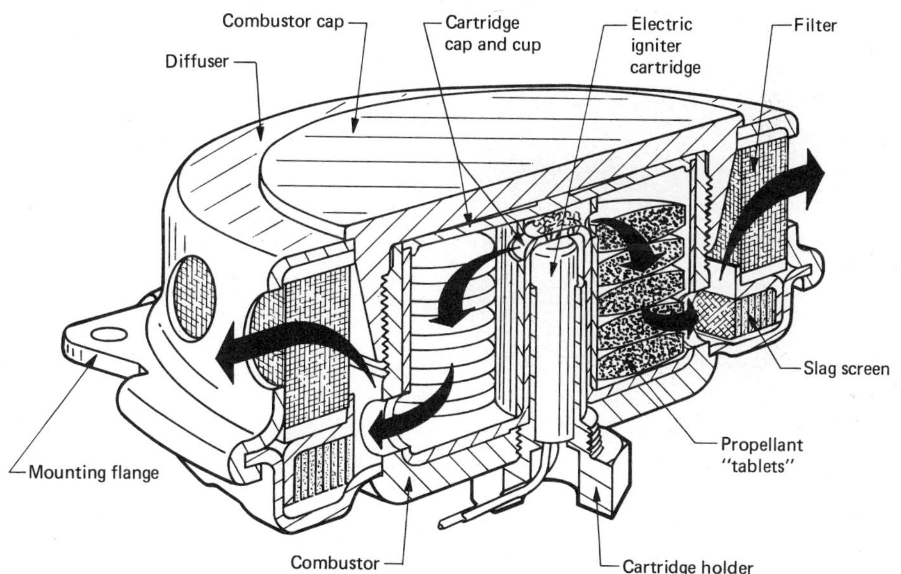


**FIG. 7-16.** Driver's air-bag inflator.

permeability to gas flow, and a contorted flow path to centrifugally separate both solid and liquid particles from the gas. The filters also have high heat capacity for cooling the gas before entering the air bag.

In addition to the driver's air bag and inflator, which are mounted in the steering wheel, there is also a much larger air bag, which is concealed in the instrument panel in front of the passengers. In small, compact automobiles, multiple inflators of the type used in the driver's steering wheel can be utilized. In larger cars, however, it is much simpler to provide a single, larger inflator as shown in Figure 7-17. The inflated volume of this air bag may be 7 cubic feet (198 liters) or greater, and the length of the inflator may vary between 18 and 36 inches (45.7 and 91.4 centimeters). This longer geometry provides greater area through which the generated gas can be delivered into

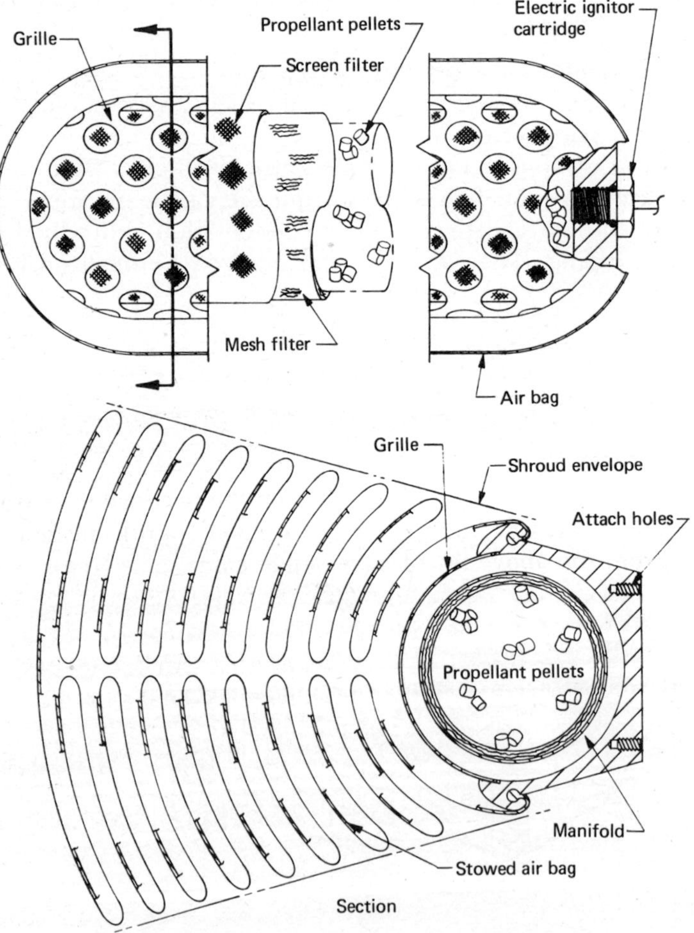


**FIG. 7-17.** Passenger's air-bag and inflator assembly.

the stowed air bag, resulting in much lower chamber pressure in the inflator. The passengers' air-bag inflator contains the same basic components as the driver's, with the main differences being their relative sizes and geometries.

So much has been written recently about the makeup of the propellants used in air-bag inflators, a brief discussion of the Department of Transportation's generated gas constituents and requirements, and how the propellant industry is meeting them, is warranted.

The air-bag passive restraint program provides a tremendous marketing opportunity for pyrotechnic gas generators, or inflators, in a nonmilitary, nonaerospace market. Very large production volumes are possible in the automotive market, permitting extensive development programs and highly cost-efficient designs. The requirements imposed on the inflators, however, are severe because they are in close proximity to humans. The safety of the design must be unimpeachable, even after years of exposure to inhospitable environments. They must withstand the most severe treatment an automobile can provide and still function when called upon to do so over a temperature range of −40 to +185°F (−40 to +85°C). Finally, it must be capable of being subsequently destroyed, along with the car, in a shredder or compactor without presenting hazards to people or equipment.

In selecting a pyrotechnic composition for the air-bag inflator, the first consideration is the composition of the gas to be generated. Due to constraints of toxicity, only a few pyrotechnically producible gases are permissible in more than trace quantities. They are:

1. Nitrogen—no limit
2. Oxygen—limited to 20 to 25 percent by flammability considerations
3. Carbon dioxide—limited to about 20 percent for physiological reasons
4. Water—limited to about 30 percent

Additionally, there are stringent limitations on the production of trace gases, such as carbon monoxide (500 parts per million [ppm]), nitric oxides (25 ppm), etc. From the above list of permissible gases, nitrogen is seen as the key member. No combination of the others will suffice without a significant addition of nitrogen.

Because the air bag must be completely filled in approximately 40 milliseconds, the pyrotechnic composition must have a higher linear burning rate in order to yield propellant "grains," or granules, of manageable shape and size. Practically, the propellant formulation must have a burning rate on the order of 0.5 to 1.5 inches (12.7 to 38.1 millimeters) per second at the selected operating pressure, which must be no more than 2,000 to 3,000 pounds per square inch (141 to 211 kilograms per square centimeter) for structural/weight reasons.

The composition must be formable into some easily producible shape, with a characteristic thickness determined by the burning rate. The most commonly used approach is powder compaction (pressing) into hollow or solid cylinders and flat "pills" or "tablets." The finished propellant pellets in all forms must withstand an environment of vibration and temperature cycling along with being tolerant of whatever pellet-to-pellet abrasive action they may encounter, without powdering, during their anticipated 10-year installed life in an automobile.

The most popular gas-generating pyrotechnic compositions for air bags are based on sodium azide ($NaN_3$). The sodium azide is typically mixed with a metal oxidizer such as ferric oxide ($Fe_2O_3$), and sometimes a nitrate or perchlorate oxidizing salt (e.g., $NaNO_3$, $KClO_4$) to enhance the burning rate. A typical reaction equation would be:

$$29NaN_3 + 4Fe_2O_3 + NaNO_3 \rightarrow 15Na_2O + 8Fe + 44N_2$$

This reaction is cool-burning (300 calories per gram), has a high burning rate (1 inch or 25.4 millimeters per second at a pressure of 1,000 pounds per square inch or 70.3 kilograms per square centimeter), and adequate gas yield and density, and is readily compacted into cylinders or pills (with the addition of trace amounts of pressing aids) on conventional powder compacting presses.

Although the above formulation meets all the general requirements for the air-bag program, the sodium azide compound is both toxic and expensive. For this reason, there is considerable interest in alternative formulations that alleviate most of these problems. Formulations which produce a combination of the permissible gases, with 30 percent or more being nitrogen, are being studied by several companies. One of the problems of these formulations is that, where nitrogen, oxygen, and carbon dioxide are all present, it is quite difficult to simultaneously suppress the generation of both carbon monoxide and nitric oxide enough to meet the stringent requirements imposed by the air-bag governing administration. Interestingly, the technology for dealing with this problem is also to be found in the automotive industry—in methods for controlling toxic exhaust emissions.

### Augmented Cool-Gas Inflation System

The wide-body jet airliner era that began in the 1970s introduced a new requirement of providing emergency escape capability for well over 500 passengers and crew members. The Federal Aviation Administration (FAA) requires a demonstration of the capability to evacuate the maximum passenger and crew load through half of the airplane doors in not more than 90 seconds from the instant the decision is made to evacuate. In actual tests with the largest of the present airliners, the Boeing 747, as many as 540 passengers and crew members have been evacuated in less than 60 seconds. This accomplishment was made possible by the pyrotechnique of utilizing the hot, cartridge-generated gas to perform two functions in a unique inflation system.

The floor of the main passenger compartment is approximately 21 feet

(6.4 meters) above ground level. Deployable ladders and stairways from this elevation were too bulky and complex for convenient stowage. An adaptation of a playground sliding board made of rubber coated cloth that could be folded into an extremely compact package and rapidly deployed and inflated met the FAA evacuation time requirement perfectly. Figure 7-18*a* shows the right side of a Boeing 747 airliner on land with escape slides deployed from five exits on the main passenger deck and one slide deployed from the upper

(*a*)

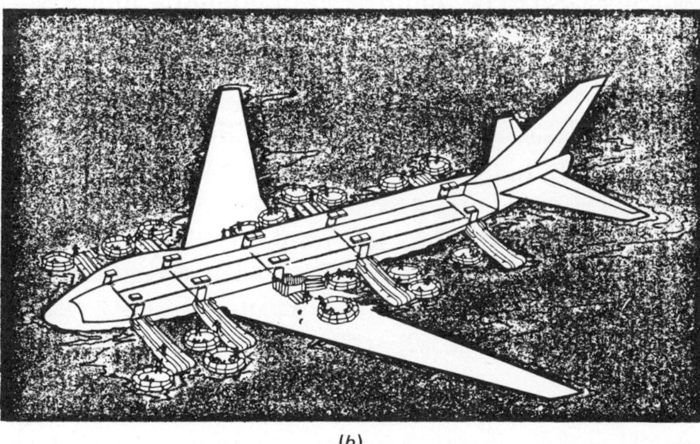

(*b*)

**FIG. 7-18.** Deployment of airliner emergency escape slides. (*a*) Land deployment. (*b*) Water deployment.

deck, which is approximately 30 feet (9.1 meters) above ground level. The left side of the aircraft has only the five slides from the main passenger deck. Figure 7-18*b* illustrates the same aircraft after having been ditched in water. Note the addition of 17 inflated life rafts. To accommodate the large number of evacuees and at the same time minimize the number of exits and slides, the aircraft doors and slides were made wide enough to accommodate two exiting evacuees at a time. At the number 3 door, located over the wing, in what appears to be a single slide from the door and "off the wing," are actually two slides. One is deployed from inside the lower liner of the door, and the "off the wing" slide is deployed from a compartment in the fairing between the wing trailing edge and the fuselage. When the aircraft is ditched in water, deployment of the off the wing slide is inhibited by a control lever on the number 3 door. To further accommodate ditching situations, the number 3 door ramp also has a gate through which the evacuees can move out onto the wing to subsequently board life rafts. Later versions of the escape slide are detachable from the aircraft and can be used for life rafts. This modification resulted in considerable weight reduction in the 747 emergency escape system even though it imposed inflation system gas-flow requirements far beyond any system available at that time. Slide volumes varied from approximately 350 to 450 cubic feet (9,912 to 12,744 liters), with a maximum allowable inflation time of 10 seconds to a final inflation pressure of 2.0 pounds per square inch (0.141 kilogram per square centimeter).

Each emergency escape slide inflation system consists of a compound gas generator composed of pyrotechnic propellant and liquid freon. Generated gas is delivered by flexible hoses to dual, two-stage aspirators (injector pumps) attached to each slide. Figure 7-19 illustrates a typical slide installation mounted to the inside of a 747 door. In normal operation the system just rides with the door each time it is opened and closed. In an emergency requiring escape slide deployment the initial position of the door would be closed. A slide deployment arming handle is accessible through an access door and is rotated to the armed position. This action engages a slide girt that is attached to the slide with a pair of brackets that are attached to the floor of the aircraft. Now the door is opened in the normal manner. After the door has manually rotated approximately 20 degrees, the door's power assist system takes over to simultaneously power the door fully open and extract the slide from the shroud. Not until the slide falls free of its shroud is the cool-gas inflation system initiated by a lanyard attaching the slide to the gas generator firing (trigger) mechanism. A safety feature prohibits deployment of the escape slide when the door is opened from the outside of the aircraft.

The inflation system gas generator is illustrated in Figure 7-20. It consists of a hermetically sealed, pyrotechnic gas-generating cartridge, encased in a high-pressure cylinder that is partially filled with liquid freon. Initiation begins when a lanyard attached to the deploying slide actuates a trigger

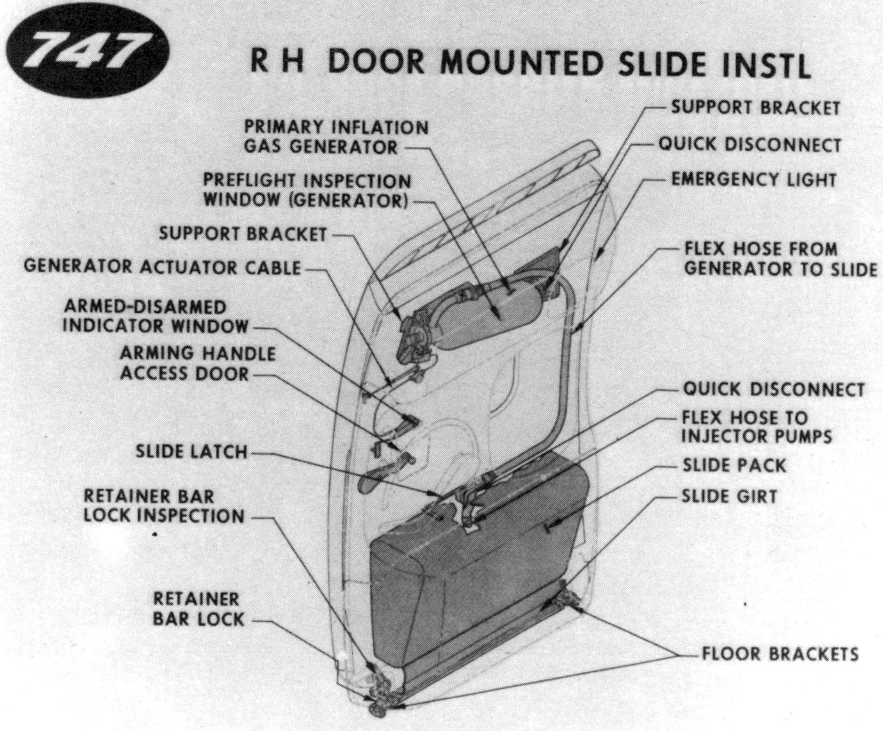


**FIG. 7-19.** Door installation of emergency escape slide.

mechanism. This mechanism consists of a spring-loaded firing pin striking a percussion primer (see Chapter 3) located at the elevated end of the propellant cartridge. The propellant burns on its exterior surface and on the cylindrical wall of a hole through its center. The hot, high-pressure gas is directed into a small plenum having two outlet orifices. One directs the gas to the outlet hose that is connected to the aspirators while the other directs the hot gas into the liquid freon, causing it to vaporize and thus create more gas that forces liquid freon through a tube located at the lowest point of the cylinder. The tube directs the liquid freon into the path of the hot propellant gases. While the liquid freon and hot propellant gas are mixing, the liquid freon vaporizes. This results in cooling the hot propellant gases, thus reducing a hazardous condition as the combination of propellant and freon gases enters the aspirators connected to the rubberized cloth slide.

Though the aspirator is not a pyrotechnic device, some understanding of its principles is necessary in order to understand the workings of the final stage of the emergency escape slide inflation system. Figure 7-21*a* is a schematic illustration of a simple aspirator consisting of a pipe located at the flared end of a short tube. When high-pressure gas is directed into the flared end of the tube from the small pipe, air in the tube is displaced from the

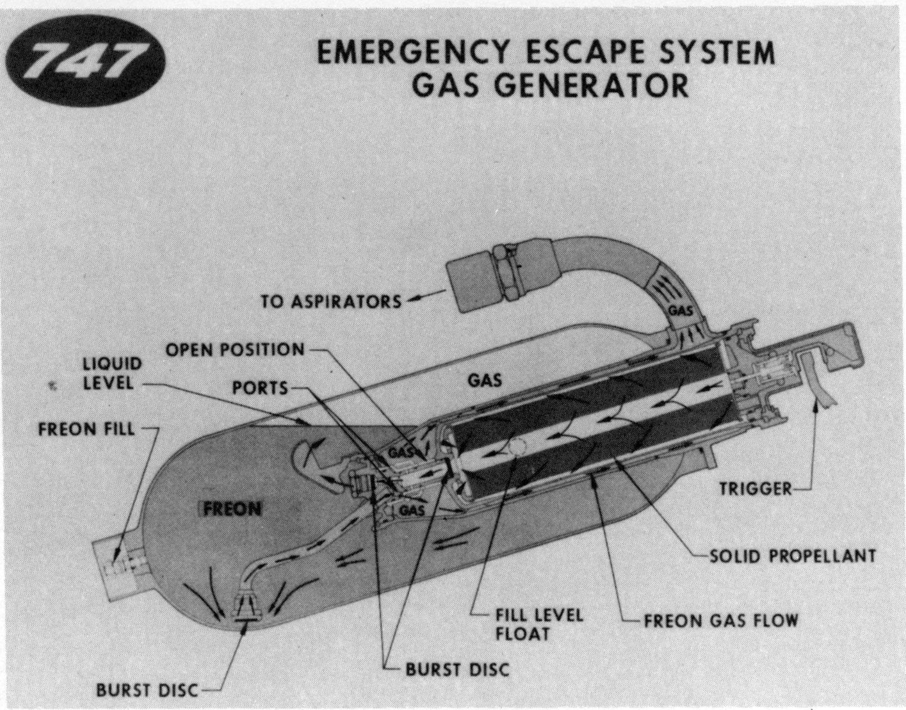


**FIG. 7-20.** Escape slide gas generator.

opposite, unflared end. This displacement creates a small pressure drop near the flared end of the tube. Ambient air then rushes into the tube to fill the partial vacuum, and it is likewise accelerated by the continuing stream of air from the pipe. The lower half of the illustration depicts the mixing of the air from the pipe with the incoming ambient air. The upper half plots the relative velocities of the gases from the center of the tube to its wall at numerous stations along the tube. As can be seen, the interrelationship between the molecular weight and velocity of the generated gas and the geometry—diameter and length—of the tube exerts tremendous influence on the total performance of the inflation system. Still another important parameter is the relative opposing pressure at the exhaust end of the aspirator tube. This pressure continues to increase as the slide is inflated. Figure 7-21*b* is a section through a simple aspirator. The pipe enters the tube through its side and makes a 90-degree bend to direct the high-pressure, cool gas down the center of the tube. Behind the elbowed pipe, near the flared end of the tube, is a butterfly check valve that folds together while ambient air is entering the flared end of the tube. When pressure in the inflating slide approaches that at the exhaust end of the aspirator tube, gas flow within the aspirator begins to reverse. To prohibit the slide from deflating, the initial reverse flow causes the two semicircular butterfly vanes

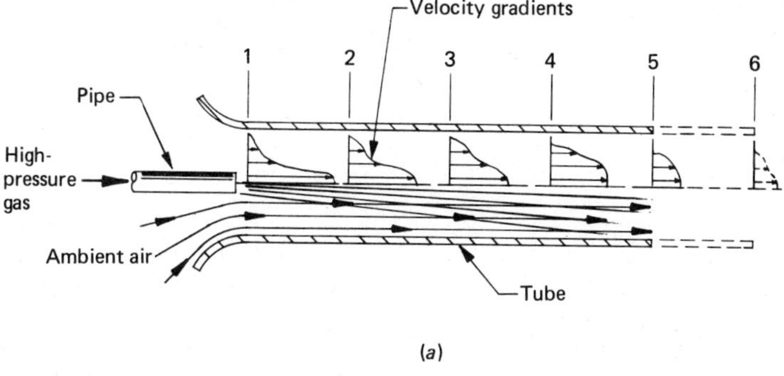


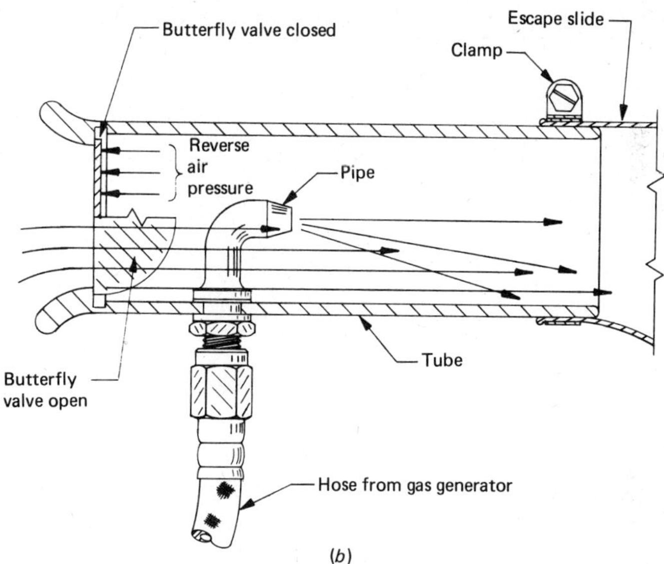


**FIG. 7-21.** Principle of simple aspirator. (*a*) Aspirator schematic. (*b*) Cutaway of aspirator.

to actuate to a closed, circular position, thus preventing a reverse gas flow from the inflated slide through the aspirator.

Two simple aspirators of this type are shown in the foreground of Figure 7-22. Included in the photograph is a whole shipset of Boeing 747 emergency escape slide inflation systems. The gas supply for the two simple aspirators is high-pressure nitrogen stored in bottles shown in front of the aspirators. The 22 additional aspirators are more complex in that they are composed of a series of concentric tubes, with the pipe being replaced with a tube shaped like a torus having perforations on the downstream side. In the background are 11 gas generators, each of which drives a pair of the compound aspirators attached to an emergency escape slide.

**FIG. 7-22.** Complete Boeing 747 complement of emergency escape slides inflation gas generators and aspirators.

## OXYGEN GENERATOR (OXYGEN CANDLE)

Ordinarily, a pyrotechnic device performs useful work only during the few milliseconds while the pyrotechnic within it is decomposing. The device may make use of the huge volume of gas generated, the gas pressure generated, the heat of decomposition, or a combination of two or more of these products of combustion. There is yet another group of pyrotechnics, the chlorates and perchlorates, whose thermal decomposition results in the release of oxygen gas and a porous ash of chloride. They have come to be known as *solid-state oxygen generators* or *oxygen candles*. The aforementioned chemical reaction is often demonstrated in freshman chemistry classes with test tubes heated over Bunsen burners. Addition of trace elements to the perchlorate will alter its decomposing characteristics; for example, a trace of manganese dioxide will cause the oxygen to be released at a much lower temperature, whereas, traces of certain metallic powders for fuel will sustain combustion after the external heat source has been removed.

The basic formula or composition needed to produce pure oxygen consists of a chlorate or perchlorate, a fuel or oxidizable material, a binder, and any purifiers necessary to produce the purity of oxygen desired. An initiation area, rich in fuel, is necessary for activation. Although there are more than a dozen oxygen-rich source chemicals, sodium chlorate is the only one commercially produced in tonnage quantities and therefore has been the most widely used compound in today's state of the art. When properly heated, sodium chlorate gives up approximately 45 percent of its weight in gaseous oxygen. Iron is the main fuel used in the body of an oxygen candle, while carbon, iron, and zirconium have been used for fuels in the initiation area which can be a separate disk or integral with the body of the candle. The reactive process may be represented by the following simplified chemical equation and reaction:

$$Fe + NaClO_3 \rightarrow FeO + NaCl + O_2 + \text{heat}$$

A binder maintains the physical integrity of the candle and aids in the smooth generation of oxygen. It also separates the particles of chlorate and distributes the heat for even thermal decomposition. Glass wool, asbestos, rock wool, and steel wool have all been used as binders. Purification can be performed by certain chemicals added to the mixture to chemically combine and to catalyze impurities to harmless compounds. Manganese and Hopcalite can catalyze carbon monoxide to carbon dioxide. Lithium oxide and barium peroxide have been used to remove any chlorine released in the thermal breakdown of the chlorides.

Normally, the oxygen candles are formed into sticks of circular or square cross sections by pressing or molding. The activation area is on one end, and since the decomposition rate is uniform along the axis, the cross-sectional area determines the oxygen flow rate. The length of the candle determines the duration of oxygen generation. It is obvious that variations in the candle's axial form can be made to generate oxygen at predictably variable flow rates. Any change in the axial cross-sectional area of the candle will result in a corresponding change in the volume of oxygen generated. Also, changes in formulation (composition) can be made to vary the rate of oxygen generation as much as 20 percent. In general, an increase in candle density reduces the rate of oxygen output. Removing the hot oxygen over the reacted area results in very little acceleration in the rate of decomposition.

The solid-state oxygen generator can be initiated by percussion primers, hot-wire initiators (see Chapter 3), hot wires that contact the activating area of the candle, or even a hand-held match. The activation area or disk may have one to four graduated layers to assure low temperature activation. Initiation can be done manually or electrically from a remote location. An electrical impulse from a pressure or temperature switch has been employed to initiate automatically an oxygen generator.

Sodium chlorate melts at 482°F (250°C), and the oxygen is released from this temperature up to 1,472°F (800°C), the melting point of sodium chloride. To ensure residue ash integrity, it is best to maintain the decomposition temperature between 482 and 1,112°F (250 to 600°C). The resultant ash normally expands slightly in cross-sectional area, although this is preceded by a slight shrinkage in the direction of the reaction. The reaction zone is a liquid, and the oxygen flows to the edges rather than through the solidified chloride ash. Heat is supplied by the oxidation of the iron fuel. The decomposition of a chlorate is also exothermic, and this heat preheats the adjacent crystals of the chlorate. Heat must be removed from the generator in order to maintain a uniform rate of decomposition. The overall heat produced is approximately 80 Btu (British thermal units) per cubic foot (2,980 joules per liter) of oxygen released.

Pressure has little affect on the decomposition rate of the chlorate if a uniform temperature is maintained at the reaction zone. Experiments have disclosed that generators cooled with a water bath give little variation in reaction rate over a range of 0.5 to 1,500 pounds per square inch (0.035 to 110 kilograms per square centimeter). The generator can develop pressures limited only by the structural capability of its container. Pressures as high as 10,000 pounds per square inch (703 kilograms per square centimeter) have been reached. The actual maximum operating pressure is readily controlled by the inclusion of a pressure relief valve built into the housing of the generator.

A small quantity of finely divided chloride vapor or aerosol is produced in the reaction zone that has to be filtered from the oxygen if it is to be used for breathing purposes. This can be done by passing the gas through a fiberglass mat or cloth. This filter can be incorporated around the candle itself or installed externally as a separate unit.

Two commercial applications of the oxygen candle are discussed in the following pages. They fall at the extreme ends of the scale for purity of oxygen generated. The first utilizes the oxygen in passengers and crew emergency breathing systems on commercial airliners, while the other application mixes the relatively impure oxygen with a gaseous fuel and is ignited as a torch to perform welding and brazing operations.

### Emergency Breathing Systems

The typical oxygen generator for commercial airline passenger cabin requirements is designed to produce oxygen at a time-varying rate which coincides with the physiological needs of passengers during descent from cruising altitude. These requirements are profiled in Figure 7-23 and are derived from Federal Aviation Administration (FAA) regulations, FAR 25.1443(4). A typical cabin altitude profile (cabin altitude pressure versus time from decompression) is plotted along with a corresponding oxygen flow requirement profile. The altitude profile is derived from the worst case rate of decompression, the emergency descent rate, and other factors influencing the timing of descent. The DC-10 utilizes one, two, and three–man oxygen generators. These three models can accommodate all possible passenger seating arrangements between aisles and in lavatories. Barometric switches,

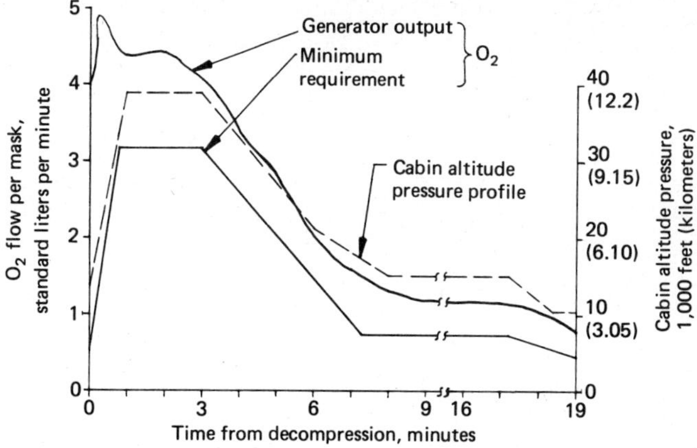


**FIG. 7-23.** Commercial airliner cabin decompression pressure profile and passenger's oxygen requirements.

located in each passenger cabin or zone, sense a drop in cabin pressure and open an overhead door, causing the deployment of the breathing masks in front and slightly above the seated position of the passengers. A short lanyard attaches the oxygen hose of each of the masks to the mechanical initiator attached to the generator. When any of the masks is pulled toward a passenger's face, the lanyard triggers the initiator, causing a flow of oxygen from the generator through the connecting hose to the passenger's breathing mask.

The generator shown in Figure 7-24*a* is the two-man model with the external heat screen removed to better present the generator's size. Figure 7-24*b* is a schematic of the percussion-initiated two-man generator. Activation begins, as stated earlier, by pulling the lanyard that releases a spring-loaded firing pin that strikes a percussion primer located near the activation end of the oxygen candle. The flame from the primer is directed through a "spit" hole onto the activator layer (fuel-rich end) of the candle whose overall length is approximately 7 inches (178 millimeters), with a burn rate approximating 0.35 inches (9 millimeters) per minute. The candle's density is approximately 0.09 pounds per cubic inch (2.4 grams per cubic centimeter). Immediately, the entire circular end surface of the candle begins to decompose. Pressure begins to build within the housing and oxygen begins to flow from the opposite end of the housing through a manifold with dual outlets and connecting hoses leading to two passenger breathing masks. A pressure relief valve is set to limit maximum internal pressure to 50 to 75 pounds per square inch (3.5 to 5.3 kilograms per square centimeter). As the hot, oxygen-rich gas travels toward the exit manifold, it passes through numerous pads, chemical filters, and refractory packing material that remove the impurities and cool the oxygen to nearly ambient temperature at the manifold outlet.

A comparable electrically initiated oxygen generator is used on the L-1011

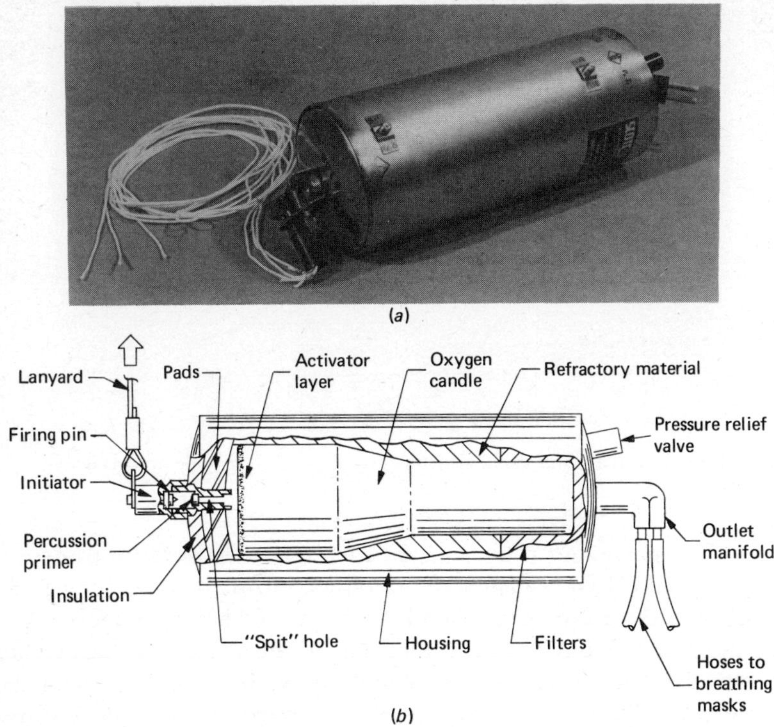


**FIG. 7-24.** DC-10 airliner percussion initiated 2-man oxygen generator. (*a*) Oxygen generator with external heat screen removed. (*b*) Schematic of 2-man oxygen generator.

airliner. It too uses one-, two-, and three-man generators; the three-man model is shown in Figure 7-25. The initiating wires enter the generator housing on the same end as the outlet manifold with the three hose fittings for the breathing masks. Internally, the wires must traverse the entire length of the housing to initiate the candle at the far end. Other internal characteristics are similar to those shown in Figure 7-24*b* schematic. Note that on the body of the housing, near the center of the heat screen, are two painted stripes—one light and one dark in color. When the oxygen generator is spent, heat on the surface of the housing causes the light-colored paint stripe to turn dark, nearly matching the color of the dark stripe. This indicator alerts attendants and maintenance personnel that this particular generator is spent and should be replaced. The third perturbation shown in the photograph is the pressure relief valve.

Constant-flow oxygen generators are on board most L-1011's as a substitute for gaseous portables used by cabin attandants for first aid situations. These units, shown in Figure 7-26, have generators which produce in excess of 4 liters of oxygen per minute for approximately 20 minutes. Spare generators can be carried, and the dispensing unit can be recharged with fresh gener-

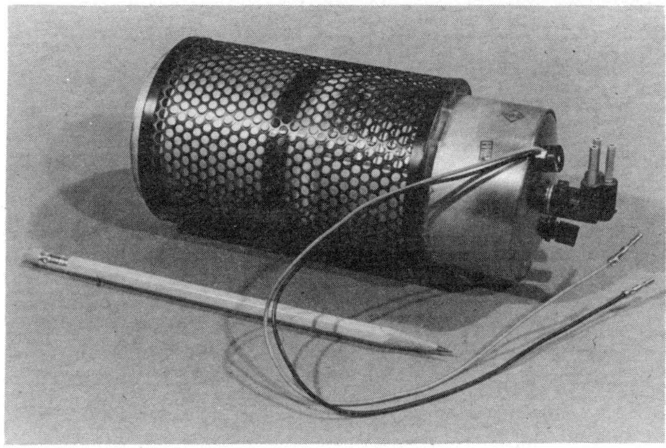

FIG. 7-25. L-1011 airliner electrically initiated 3-man oxygen generator.

ators in flight. Caution must be taken since these cartridges do not have heat screens attached and they are very hot. The screens and insulators are within the oxygen-dispensing unit.

Other adaptations of the solid-state oxygen generators or oxygen candles have been in situations involving evacuation from underground mines, escapes from below-deck fires on navy and merchant ships, and submarine emergencies. They can produce oxygen that meets United States Pharmacopoeia (USP) standards, as well as Aviators' Breathing Oxygen standards

FIG. 7-26. Portable oxygen generator with replaceable oxygen candle.

regarding physiological and toxicity requirements. They can produce oxygen when the ambient temperature is as low as −65°F (−53.9°C) or as high as 140°F (60°C). They do not contain high-pressure gas and so are free of high-pressure cylinders that must be periodically hydrostatically tested. Their service life is at least 10 years. Submarine candles, experimental units, and some generators used in breathing apparatus have functioned well after 15 to 20 years in storage. The reliability of aircraft-type solid-state generators has been demonstrated to be 99 percent, with a 90 percent confidence level.

### Welding and Brazing Torch

The elementary chemistry experiment of heating a chlorate or perchlorate in a test tube to produce oxygen, as described in the beginning of this section, was often followed by adding steel wool to the generated oxygen and igniting it with a match to illustrate how a material normally thought of as being incombustible will burn in an oxygen-rich environment. Based upon this principle is the oxyacetylene welding torch. Here, two gases, oxygen and acetylene, are stored in high-pressure bottles with pressure regulators and mixing valves at the torch handle to control the flame. Welding systems such as this require a lot of skill to operate and are very expensive to buy. The solid oxygen welding and brazing torch kit, shown in Figure 7-27*a*, has simplified the system considerably, and the cost has been reduced to where it is affordable by the hobbyist, do-it-yourself mechanic, and weekend plumber. The basic kit consists of an assembly composed of a solid-state oxygen wand, a low-pressure bottle of propane (fuel) with a flow control valve, hoses directing the oxygen and propane to a mixing plenum within the handle of the operator's torch, a flint striker for lighting the flame, brazing rods, solid oxygen sticks with ignition squares, and an instruction manual. Three color-coded groups of oxygen sticks are all the same size and generate approximately the same volume of oxygen. The differences are in their respective binders, which alter their individual burn times, i.e., red–8 minutes, white–10 minutes, and blue–12 minutes. At the torch, the higher oxygen flow rates generated by the red and white sticks require correspondingly larger quantities of propane fuel than do the blue, slower-burning sticks. The net result: the faster the burning rate of the oxygen stick, the larger the flame at the torch. It's important to note that the flames produced with the red and white oxygen sticks are larger, not hotter, than those produced by the blue sticks. Relatively larger and smaller flames are required to braze correspondingly thicker and thinner materials.

Figure 7-27*b* is an exploded view of the torch assembly. Operation of the torch is as follows. The instruction manual advises the operator which oxygen stick and brazing rod should be used; the decision is based primarily upon the thickness of the materials to be joined. An igniter square is placed in the base of the oxygen tray with the proper oxygen stick placed on top. With a match, the operator lights a corner of the igniter. When the oxygen stick shows evidence of burning, the oxygen wand is inserted into the canister.

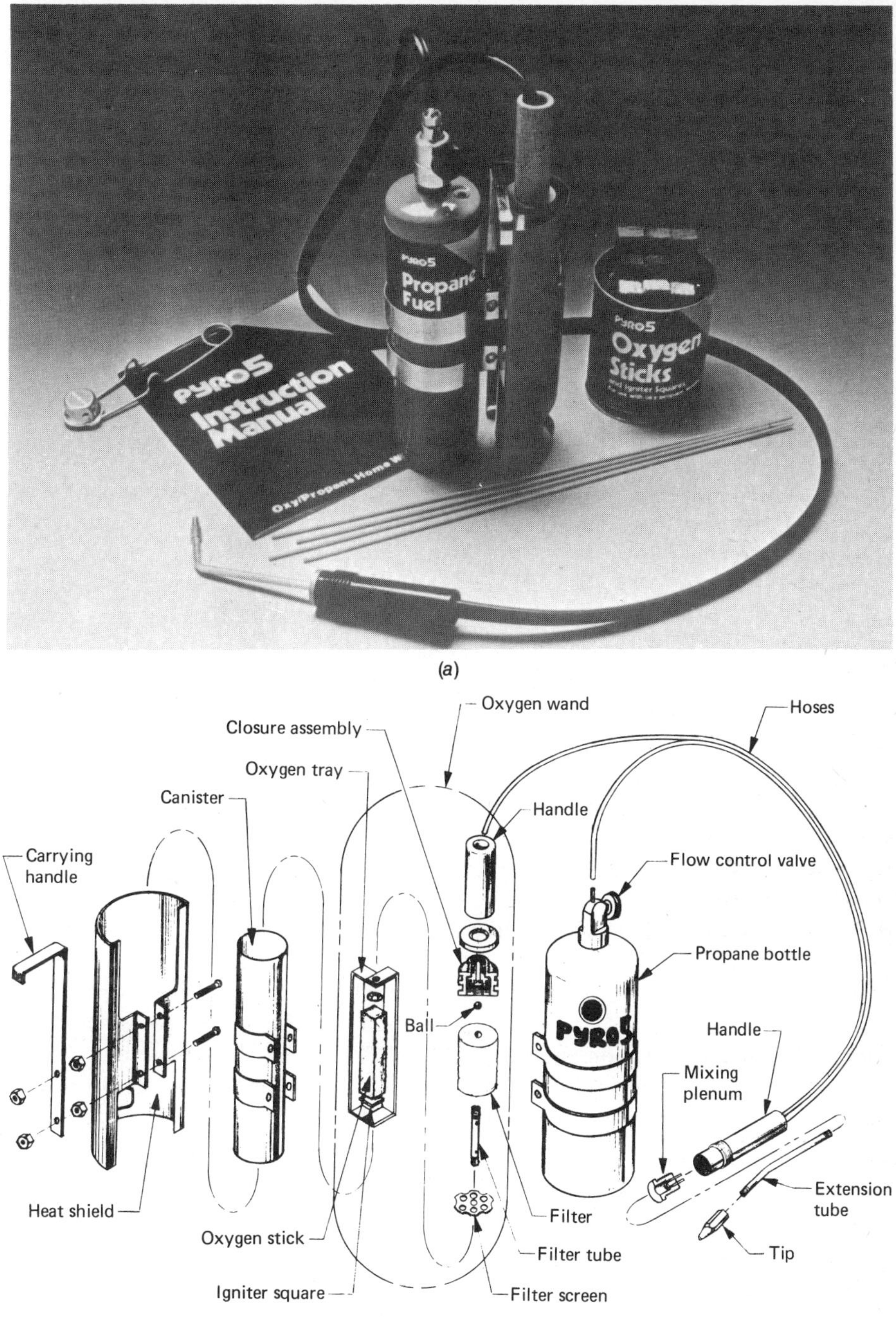

(a)

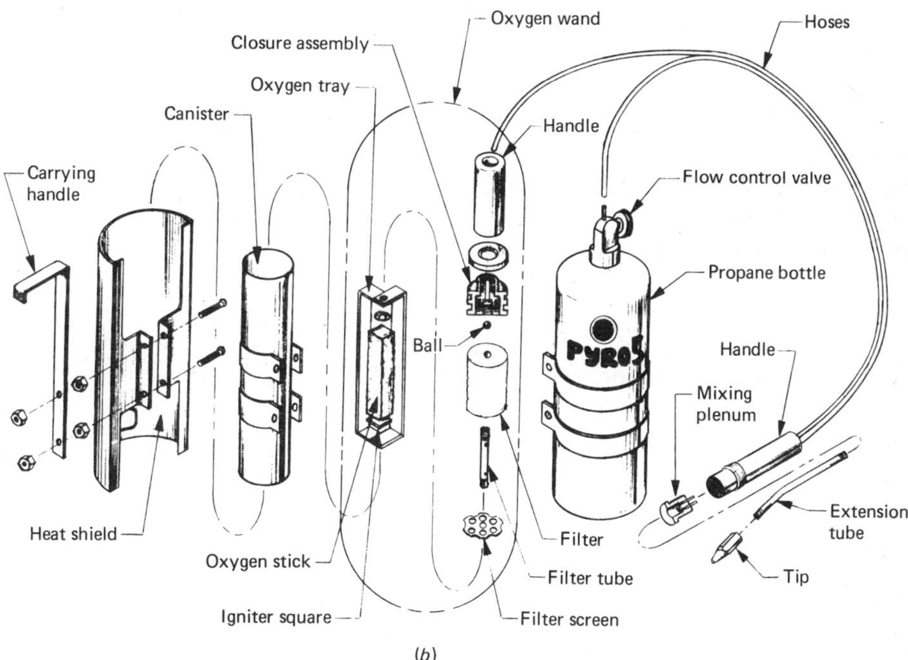


(b)

**FIG. 7-27.** Solid oxygen welding and brazing torch. (*a*) Basic kit. (*b*) Exploded view of torch assembly.

The flow control valve atop the propane bottle is opened slightly, and the outlet end of the torch is directed into the cup of the flint striker as the striker is actuated. Sparks from the flint will ignite the oxygen-propane mixture. The propane flow control valve is then adjusted until an outer blue-white flame with a bright blue inner cone approximately 0.25 to 0.5 inch (6.3 to 12.7 millimeters) long is achieved. This is when the flame attains its highest temperature of approximately 5,000°F (2,760°C).

Other torch characteristics include a filter within the oxygen wand that prevents residue ash particles from entering the filter tube through holes in its side. The lower end of the filter tube is closed. A small ball rests atop the opened end of the filter tube inside the closure assembly. A minimal amount of oxygen pressure is required to unseat the ball from the upper end of the filter tube, thus allowing low-pressure oxygen to enter the transfer hose to the mixer plenum in the handle of the torch. In the unlikely event that excessive pressure is generated within the oxygen canister, the entire oxygen wand will displace vertically until the O ring on the closure assembly enters the flared section of the oxygen canister, thus allowing the excess oxygen to escape harmlessly into the atmosphere. Then the oxygen wand can be reseated in the canister. The instruction manual concludes with illustrated examples of numerous types of elementary welding and brazing procedures and applications.

chapter 8

# Quality Assurance and Control

The unique characteristic of pyrotechnics—that of being consumed during their short useful life—has imposed stringent requirements in their development, qualification, and manufacturing procedures. Performance and safety margins are demonstrated to assure reliability even when a substantial number of manufacturing variables combine to degrade or enhance the performance characteristics of the device.

## DEVELOPMENT TESTING

The initial phase of development testing of a pyrotechnic device commences with a configuration arrived at by analysis, experience, and judgment. Since pyrotechnics are "one-shot" devices, the development "boiler-plate" device is designed to be refurbishable for reuse with a minimum number of replacement parts. Cartridge bodies being designed to accept as much as 20 percent more propellant than initially calculated will save considerable time if the first development firing test should not perform the required pyrotechnic function.

Subsequent development test firings will be conducted to determine the minimum quantity of pyrotechnic material (usually by weight) required to perform the function. This minimum quantity plus some small added amount, usually 5 to 8 percent, establishes the minimum tolerance of the quantity of pyrotechnic material to be used in subsequently manufactured cartridges. Since most initial development tests are conducted at ambient temperatures, the additional charge slightly more than compensates for the reduction in the pyrotechnic performance at low temperatures—most often −65°F (−54°C).

Development testing generally culminates in a series of design verification tests (DVT), utilizing hardware of near production configuration, that informally demonstrates the capability of the newly developed pyrotechnic device to satisfactorily meet the more severe performance requirements. These often include combinations of high- and low-temperature firings coupled with high- and low-tolerance pyrotechnic loadings and single and dual cartridges. They provide pressure versus time instrumentation data from representative cartridges fired in "bombs" of known volume to establish a performance basis for destructive lot acceptance tests (DLAT) of subsequent production lots. Although the pressures generated in these instrumented "bombs" are generally lower than those developed in the functioning pyrotechnic device, due to greater free volume, they do present the performance characteristics of the cartridge in that particular "bomb."

## QUALIFICATION TESTING

This series of tests formally demonstrates that the pyrotechnic device is capable of meeting all performance requirements under all simulated operating conditions. These conditions include environments such as exposure to sand and dust, moisture with and without salt spray, dynamic shock, temperature and pressure cycling concluding in most cases with a firing test at high, low, or ambient temperatures, and pressure. Cartridges utilized in these tests may be loaded with a nominal, underload, or overload quantity of pyrotechnic material. The underloaded cartridges are used singularly in the device, with an inert plug in the opposite cartridge port to demonstrate performance with an abnormally small pyrotechnic charge. The overloaded cartridges are used in pairs within the device to demonstrate structural safety margins as well as performance. The nominally loaded cartridges are fired singularly and in pairs in the device after having been exposed to the aforementioned environments.

Although all qualification test programs require the pyrotechnic devices to be exposed to their anticipated environments prior to being fired, the best method of environmental application is yet to be agreed upon by pyrotechnic quality control and assurance engineers. The two most frequently used methods are known as the *drop-out* and the *sequential* methods of qualification testing.

### Drop-Out Qualification Testing

Table 8-1 shows the drop-out method of qualification testing. In general this method sequentially exposes all of the pyrotechnic devices to a predetermined sequence of environments. At the completion of each environmental exposure a small number of the devices are withdrawn and fired at high, ambient, and/or low temperature. The main argument in favor of this method is that should a firing test failure occur, it is relatively easy to ascertain a design weakness within the pyrotechnic device. Counterarguments say the last environment alone may have been able to cause the failure or that a different sequence of environmental application might have caused the pyrotechnic device to fail earlier. Theoretically, whenever a failure occurs during a qualification test, the test stops, the weaknesses within the devices are corrected, and the tests begin all over again. As can be seen, unless all the remaining pyrotechnic devices are discarded and replaced by corrected new

**Table 8-1.** Drop-Out Qualification Testing

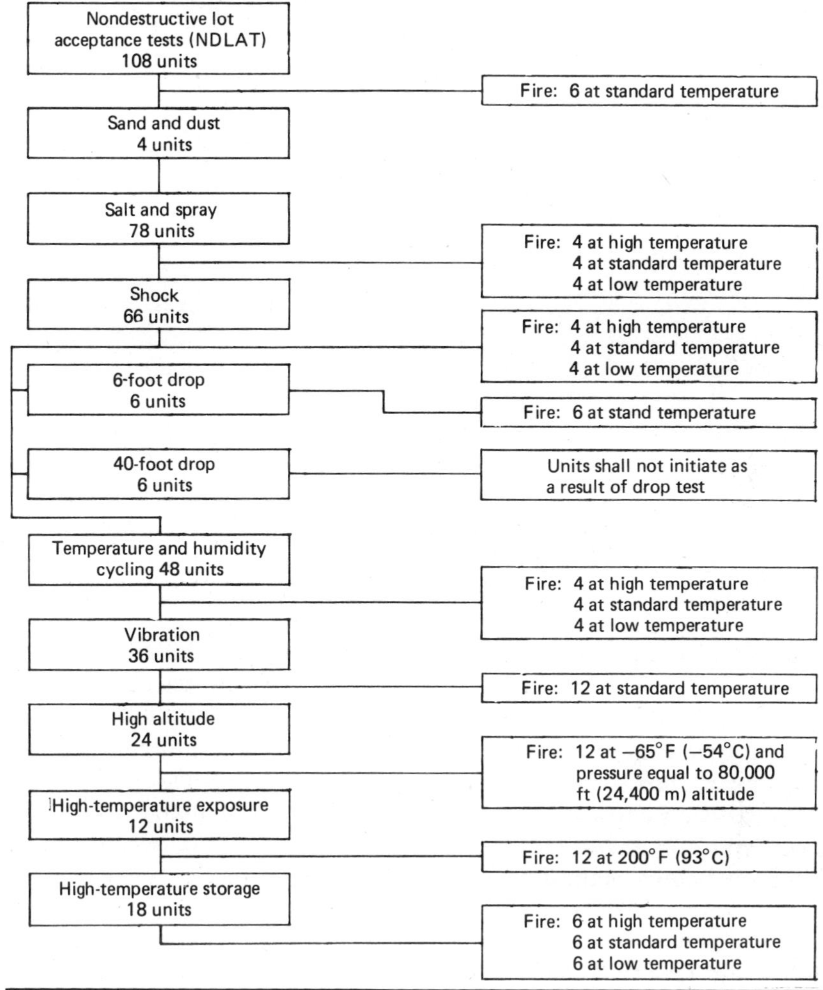


devices, a large portion of the devices are going to be exposed to the prefailure environments more than once. This overtesting alone could cause or contribute to a subsequent unrealistic failure. If all new devices, with weaknesses corrected, are used to replace the entire quantity of qualification test specimens, it can be understood how such a failure(s) could astronomically increase the cost of the devices.

## Sequential Qualification Testing

An alternative method of qualification testing is the sequential method, which is depicted in Table 8-2. The sequence in which the environments are applied to the pyrotechnic devices more closely follow those the devices will experi-

**Table 8-2.** Sequential Qualification Testing

| Test phase | Title of test | A | B | C | D | E | F | G |
|---|---|---|---|---|---|---|---|---|
| | Test group | A | B | C | D | E | F | G |
| | Test quantity | 12 | 18 | 18 | 1 | 1 | 3 | 3 |
| | | Test sequence | | | | | | |
| Check out | NDLAT | 1 | 1 | 1 | 1 | 1 | 1 | 1 |
| Climatic | Humidity | | 2 | 2 | | | | |
| | Salt fog | | 3 | 3 | | | | |
| Transportation | Random vibration | | 4 | 4 | | | | |
| Mission | Transient vibration | | 5 | 5 | | | | |
| | Vacuum/temperature cycling | | 6 | 6 | | | | |
| | High-temperature exposure +350°F (212°C) | | 7 | 7 | | | | |
| | Shock | | 8 | 8 | | | | |
| Off limit | 8-foot (2.4 meters) drop | | | | | | 2 | |
| | 40-foot (13.6 meters) drop | | | | | | | 2 |
| Performance firing test | Ambient temperature +70°F (+56.6°C) | 2 | | | | | 3 | |
| | High temperature +350°F (212°C) | | 9 | | | | | |
| | Low temperature −65°F (−54°C) | | | 9 | | | | |
| | Locked shut +350°F (212°C) | | | | 2 | | | |
| | Locked shut −65°F (−54°C) | | | | | 2 | | |

ence in actual usage. Since the devices may be utilized in numerous applications, it follows that the sequence of environmental applications should be varied. Except for some of the initial, less severe environments, the devices are generally exposed to the entire sequence of environments with subsequent firing of one group at a time. The obvious advantage of this method is that should a failure occur in a group, only the remaining devices in that group have to be discarded and replaced. After the weakness(es) have been rectified, the testing begins again with all new devices.

More often than not, in both the drop-out and sequential qualification test methods, when failures occur, an in-depth failure analysis is initiated to ascertain the cause. A design deficiency within the pyrotechnic device is seldom the cause of the failure if qualification testing has been preceded by adequate development and design verification test programs. Faulty instrumentation and/or improperly designed test fixtures can also be the cause of apparent or real qualification test failures.

## LOT ACCEPTANCE TESTING

Production lots of pyrotechnic devices manufactured after the completion of qualification testing must undergo a two-phase series of tests to be certified that they will subsequently meet the same performance requirements demonstrated by a similar lot, manufactured under the same processes and procedures, during qualification testing. The combined series of tests are known as lot acceptance tests, and the two phases are nondestructive lot acceptance tests (NDLAT) and destructive lot acceptance tests (DLAT). Each pyrotechnic device within the lot is traceable by part number, lot number, and individual serial number. This extensive traceability program is both time-consuming and expensive. If, however, a problem should arise with any pyrotechnic device of a particular lot subsequent to lot certification, all remaining suspected devices from that lot are readily accountable. Those devices already consumed can be disregarded, and those remaining in the suspected lot can either be destroyed or reworked and subsequently recertified. Without traceability, every device from all previous lots would be suspect.

### Nondestructive Lot Acceptance Tests (NDLAT)

As the name implies, the NDLAT are a series of inspections and tests that may or may not be imposed on each individual device in the entire lot. The specification for the pyrotechnic device will stipulate which NDLAT are to be imposed on all or a portion (usually expressed in percent) of the total lot quantity. A summary of the NDLAT performed on a typical lot of pyrotechnic devices is as follows:

**Examination of Product** All component and assembly manufacturing records are reviewed to verify that materials, pyrotechnic charge(s), design, construction, dimensions, workmanship, and markings comply with all applicable specifications and drawings.

**Leakage Test** The purpose of this test is to verify the hermetic seal of the pyrotechnic device. Helium is generally used because of its small molecular size. Leak rates as low as one one-millionth of a cubic centimeter per second are readily detectable.

**Radiography Test** Two types of radiographic inspection techniques are utilized with pyrotechnic devices: X-ray and N-ray (neutron ray). Both types of radiography are generally supplementary to each other rather than competitive. Through some materials an X-ray and an N-ray will look identical. Through others, such as lead, rubber, and porous materials, details would be completely obliterated on an X-ray film, whereas an N-ray would clarify all details within the device. Some of the defects that are detectable via radiographic inspection include missing, partial, and unconsolidated pyrotechnic charges; voids and gaps within and between charges; charge density variations; foreign matter within pyrotechnic charges; out-of-position charges and components; missing components such as O rings, spacers, and washers; and excessive materials such as solder, flux, and adhesive. The primary limitation of radiographic inspection is not always so obvious to the inexperienced viewer. The films do present a graphic picture of the internal condition of the pyrotechnic device at the moment the picture was taken. Even then, interpretation of the film requires years of experience to differentiate between a defect in the device and a blemish in the negative. Only careful handling and storage after radiography can assure that quality pyrotechnic devices will retain their initial reliability intact until fired.

**Verification of Electrical Characteristics** Bridgewire and insulation resistance tests are conducted in accordance with detailed specification tolerances. These specifications vary with each type of electroexplosive device (EED) and therefore will not be reviewed here.

## Destructive Lot Acceptance Tests (DLAT)

The DLAT is a performance firing test conducted on a predetermined sample size picked at random from the entire lot of pyrotechnic devices. A typical lot size versus sample size is shown in Table 8-3. Economic advantages are realized if single large lots of pyrotechnic devices can be purchased rather than several small lots. The main benefit is immediately forthcoming in the proportionately smaller number of DLAT samples fired from large lots as compared to several smaller lots equal to the larger lot. For example, referring to Table 8-3, the DLAT sample size for a single 2,000-unit lot is 95, whereas the total sample size for four 500-unit lots is a total of 188. There are 93 more deliverable units with the single large lot. In addition, suppliers bidding on larger lots can offer comparably lower unit prices by amortizing expensive, sophisticated tooling over a greater number of units.

If the device is so designed to have separable pyrotechnic cartridges, the cartridges are submitted to a series of instrumented "bomb" tests, as initially conducted in the design verification tests (DVT) discussed earlier in this chapter. The instrumented pressure-time performance data of the lot sample are compared with those of the DVT data. Pass/fail performance criteria are established in advance of the DLAT firings.

If, however, the pyrotechnics are inseparable from the device, for example, a guillotine, frangible bolt, release nut, then performance criteria relative to the total assembly are required. This may include a guillotine severing a maximum-size umbilical sample; a frangible bolt separating while under some adverse load condition; or a release nut functioning under cryogenic temperature combined with maximum load on a mating stud.

**Table 8-3.** Destructive Lot Acceptance Test Sample Size

| Lot size* | Sample size† |
|---|---|
| 11-90 | 10 |
| 91-150 | 10 plus 11.7% of quantity above 90 |
| 151-280 | 17 plus 9.2% of quantity above 150 |
| 281-500 | 29 plus 8.2% of quantity above 280 |
| 501-1,200 | 47 plus 4.3% of quantity above 500 |
| 1,201-3,200 | 77 plus 2.3% of quantity above 1,200 |

*Lot size equals sample size plus usable units.

†Fractional sample sizes 0.5 and above shall be rounded upwards and sizes below 0.5 shall be rounded downward.

DLAT of detonators is usually done by supporting the detonator in a test fixture within a prescribed distance from a small metal block at the time of initiation. The detonator will impart an impression or dent in the test block whose characteristics, such as material, heat treatment, and hardness, have been carefully controlled. The pass/fail criterion of the detonator sample is a minimum prescribed indentation in each of the blocks.

In DLAT, shaped charges are tested by firing them into controlled targets with pass/fail criteria being either minimum specified target penetration or minimum specified target thickness severed. Another DLAT associated with linear explosives is a minimum detonation velocity. Break wires are attached, at a known distance apart, to a sample of linear explosive. The opposite ends of the break wires are attached to time-recording instrumentation. A simple calculation involving the distance between the two break wires and the time it took for the detonation wave to travel this distance can determine the detonation velocity of the test specimen.

### Lot Acceptance Data Packages

Pertinent records and data associated with a lot of pyrotechnic devices are accumulated and submitted with the lot at the time of acceptance testing. Some of the data pertain to the whole lot in general, while others are applicable to individual serialized devices within the lot. Examples of the types of data included in a data package are as follows:

1. Certified acceptance reports, including the date of manufacture of the pyrotechnic devices, the lot and serial number of the initiator(s) installed, and the lot number of the pyrotechnic material(s) used.
2. Certified list of all detail parts by numbers and their applicable drawings, including inspection serial numbers for the entire lot.
3. Copies of all failure reports and associated corrective action records.
4. Documented final inspection records, including a set of radiographic negatives of each pyrotechnic device in the lot and a copy of the radiographic certification prepared by the performing vendor.
5. A copy of the pressure-time curves of the DLAT firings, dent block impression depth, and other performance parameters which may include detonation velocity, delay times, etc.
6. Material tensile strength test results, requirements, proof loading or proof pressure test results, and accompanying performing vendor's certificate.
7. Statement certifying the formula for the pyrotechnic charge(s) is the same as that used for the manufacture of the qualification lot as well as caloric test data of a sample from the current pyrotechnic material(s) lot or batch.
8. Copies of all NDLAT data on each serialized device, such as hermetic seal test data, bridgewire and insulation resistance test results.

**9.** A copy of the pyrotechnic devices explosive classification obtained from the Interstate Commerce Commission (ICC).

### Lot Certification

Upon successful completion of all lot acceptance tests, a lot certificate is issued by the manufacturer that identifies the pyrotechnic device assembly number, name, lot number, and the serial number of all the devices being certified.

## PREFLIGHT VERIFICATION TEST (PVT)

Subsequent to the successful completion of the lot acceptance tests it has become standard operating procedure in aerospace applications to conduct an annual preflight verification test (PVT) on all lots of pyrotechnic devices. A DLAT is performed on one unit from each flight-certified lot of each device to assure that no deterioration from handling, shipment, or storage has occurred which could result in unacceptable performance. Lots that have successfully passed the PVT are scheduled for flight installation during the following year. The criteria for performance are the same as for DLAT.

## MANDATORY INSPECTION POINTS (MIPS)

Within the pyrotechnic industry's quality assurance programs there has evolved a system of mandatory inspections or in-process tests at critical stages of manufacture. Their necessity is not always obvious. In fact, most have come into being after a series of pyrotechnic device failures that occurred after the regular series of qualification tests had been successfully completed. If the inspection is necessary on every device in the lot and it should happen to be time-consuming, the resolution may either be more accurate tooling, modification of the fabrication or assembly sequence or process, or a combination of all three. If the device should be a high-volume unit, a design change may even be in order. The latter may then necessitate a delta or mini-qualification test to verify that the changes have not altered the device's capability to meet its performance requirements.

# Index

# Index


Air-bag inflators, 155–159
  illustrations, 157, 158
Air-bag restraint system, 151–160
  illustrations, 152–158
Airliner emergency escape system, 160–166
  illustrations, 161, 163–166
Apollo CSM guillotine, 29–30
  illustration, 29
Aspirator, 163–166
  illustrations, 165, 166
Association of American Railroads, Bureau of Explosives, 148
Automobile passive restraint systems, 150–151

Bacon, Roger, 5
Ballistic hot-gas-energized initiator, 18–19
  illustration, 18
Battery-fired connectors, 120, 122–127
  illustrations, 122–123, 125, 126
Black powder (gunpowder), 5, 88
Blasting cap, 11
  illustration, 12
Boyle, Robert, 144
Brazing torch, 172–174
  illustration, 173
Breathing systems, emergency, 168–172
  illustrations, 169–171
Bridgewire, 8, 12–15
  illustrations, 12–14
Brisance, 24
Buttress threads, 79

Cartridge-actuated devices, 75–127
  advantages, 76–77
  for diesel engines (*see* Diesel engines, cartridge-starting systems for)
  electric utility products (*see* Electric utility products)
  pin pullers, 108–109
    illustration, 109
  powder-actuated fastening systems (*see* Powder-actuated fastening systems)
  retractor, 106–108
    illustration, 106
  separation/release devices (*see* Separation/release devices)
  switches, 109–111
    illustration, 110
  thrusters, 103–106
    illustration, 104
  valves, 111–113
    illustration, 112
CDC (confined detonating cord), 39, 40
  illustration, 39
"Chinese snow," 5

Confined detonating cord (CDC), 39, 40
illustration, 39
Conical shaped charges (CSC), 47–53
illustrations, 48, 50, 51, 53
table, 47
Connectors, 119–127
battery-fired, 120, 122–127
illustrations, 122, 123, 125, 126
internally-fired, 119
percussion-fired lugs, 120
illustrations, 121–122
underground distribution splice, 125–127
illustration, 126
Cool-gas inflation system, augmented, 160–165
illustrations, 161, 163–166
Cord, detonating, 40
Crew escape systems, 137–142
illustrations, 138, 139, 142, 143
Cyclonite (Herogen; RDX), 7

Deflagrating propellants, 8
Deflagration, 6
Demolition of bridges, 57–59
illustrations, 58–60
Detonating cord, 40
confined, 39–40
mild, 29–34
shielded mild, 34–39
Detonating cord initiator, 17, 18
illustration, 18
Detonation, 6
Detonators, 22–23
illustration, 23
Destructive lot acceptance tests (DLAT), 180–181
Development testing, 175–176
Diesel engines, cartridge-starting systems for, 94–103
cranking performance, 101–103
illustrations, 95, 101, 103
starting system, 95–97
illustrations, 96
system components, 97–101
illustrations, 98, 100
Drop-out qualification testing, 176–177
table, 177

Electric initiators, 11–15
illustrations, 12–14
Electric utility products, 113–127
battery-fired connectors, 120, 122–127
illustrations, 121–123, 125, 126
class 1 and 3 connectors, 123–125
illustrations, 123, 125
internally-fired connectors, 119–120
line taps, 113–119
illustrations, 114–116, 118, 119
percussion-fired lugs, 120
illustrations, 121, 122
underground distribution splice, 125–127
illustration, 126
Energy transfer system (ETS), 140–142
Escape systems:
airline, 160–166
illustrations, 161, 163–166
crew, 137–142
illustrations, 138, 139, 142, 143
slides, 161–162
illustration, 163
Expanding tube assembly (XTA), 33–34
illustration, 33
Exploding bridgewire (EBW) initiator, 14–15
illustration, 14
Explosion, 5–6
Explosive-energized initiator, 15–18
illustrations, 16–18
Explosive valves, 111–113
normally closed, 112–113
illustration, 112
normally open, 111
illustration, 112
Explosives:
classification, 6
high (secondary), 6–7
linear, 27–29
illustration, 28
liquid, 7, 9, 50
low (propellant), 6, 8

Fairburn, William A., 145
Fastening systems, powder-actuated (*see* Powder-actuated fastening systems)
Federal Aviation Administration (FAA), 160, 168
Federal Motor Vehicle Safety Standard 208, 150
Flares (fusees), 147–148
  illustration, 147
Flashbulb, 134–136
  illustration, 135
Flexible linear shaped charge (FLSC), 59–73
  illustrations, 63, 66, 67, 69, 70, 72
  tables, 62, 65
Forcible-entry hole cutters, 64–68
  illustrations, 66, 67
  table, 65
Frangible bolts, 79–84
  illustrations, 81–83
Frangible links, 84
  illustration, 85
Frangible nuts, 77–79
  illustrations, 77, 80
Friction initiated devices, 142–148
  fusees (flares), 147–148
    illustration, 147
  matches, 144–147
    illustration, 146
Fusees (flares), 147–148
  illustration, 147

"Greek fire," 5
Guillotine, Apollo CSM, 29–30
  illustration, 29
Gunpowder (black powder), 5, 88

Herogen (cyclonite; RDX), 7
Hexanitrostilbene (HNS), 7
High explosives, 6–7, 27
HNS (hexanitrostilbene), 7
Hot patches, 148–150
  illustration, 149
Hot-wire initiator, 13–14
  illustration, 13
Hydrostatic initiator, 20–21
  illustration, 21

Inflation systems, 150–165
  augmented cool-gas, 160–165
    illustrations, 161, 163–166
  automatic air-bag restraint system, 151–160
    illustrations, 152–158
  automatic belt-restraint system, 151
  automobile passive restraint systems, 150–151
Initiation power sources and transfer media, 11
  illustration, 12
Initiators:
  ballistic hot-gas-energized, 18–19
    illustration, 18
  blasting cap, 11
    illustration, 12
  detonating cord, 17, 18
    illustration, 18
  electric, 11–15
    illustrations, 12–14
  exploding bridgewire (EBW), 14–15
    illustration, 14
  explosive-energized, 15
    illustrations, 16–18
  friction, 142–148
    fusees (flares), 147–148
    illustration, 147
    matches, 144–147
    illustration, 146
  hot-wire, 13–14
    illustration, 13
  hydrostatic, 20–21
    illustration, 21
  laser-energized, 21–22
    illustration, 22
  manual, 19–20
    illustration, 19
  mechanical-energized, 19–21
    illustrations, 19, 21
  SMDC, 16
  thru-bulkhead (TBI), 17
    illustration, 17
Interstate Commerce Commission (ICC), 182

Jet-Axe, 64–68
  illustrations, 66, 67
  table, 65

Laser-energized initiator, 21–22
  illustration, 22
Leakage test, 179
Line taps, 113–119
  illustration, 114–116, 118, 119
Linear explosives, 27–28
  illustration, 28
Linear shaped charge (LSC), 53–59
  illustrations, 56, 58–61
  table, 55
Liquid explosives, 7, 9, 50
Lot acceptance data packages, 181–182
Lot acceptance testing, 178–182
Low explosives, 6

Manual initiator, 19–20
  illustration, 19
Matches, 144–147
  illustration, 146
Mechanical-energized initiator, 19–21
  illustrations, 19, 21
Micro-balloons, 64
Mild detonating cord (MDC), 29–34
  illustrations, 29–33
Monroe, Charles E., 42
Monroe effect, 42

National Safety Council, 148
Nitroglycerin, 7
Noble, Alfred, 7
Nondestructive lot acceptance tests (NDLAT), 179

Ordnance, 2
Oxygen generator (oxygen candle), 166–168
  emergency breathing systems, 168–172
    illustrations, 169–171
  welding and brazing torch, 172–174
    illustration, 173

Pascal's law, 33
Penetration of shaped charges, 44–46
Pentaerythritol tetranitrate (PETN), 7
Percussion-fired lugs, 120
  illustrations, 121, 122
Percussion primers, 23
  illustration, 24
PETN (pentaerythritol tetranitrate), 7
Phossy jaw, 145
Photoflash bulb, 134–136
  illustration, 135
Picric acid, 6
Pin-pullers, 108–109
  illustration, 109
Powder-actuated fastening systems, 88–94
  base material, limitation of, 94
  fastener holding power, 92–94
    in concrete, 92–94
      illustration, 93
    in steel, 94
      illustration, 93
  fasteners, 91–92
    illustration, 91
  low-velocity, indirect acting system, 87–89
    illustration, 88
  power loads, 89–90
    table, 90
  standard-velocity, direct acting systems, 89
    illustration, 88
Primacord (detonating cord), 40
Primers, 23–25
  percussion, 23–24
    illustration, 24
  stab, 24–25
    illustration, 24
Priming materials, 8
Propellants, 6, 8
Pusey, Joshua, 145
"Pyro-Techniques," 3
Pyrotechnic photoflash bulb, 134–136
  illustration, 135
Pyrotechnics:
  advantages of, 2
  definition, 2
  devices on Apollo, Gemini, and Mercury spacecraft, 2
Pyrotechnology, 3

Qualification testing, 176 – 178
drop-out, 176 – 177
table, 177
sequential, 177 – 178
table, 178
Quality assurance and control:
development testing, 175 – 176
lot acceptance testing, 178 – 182
mandatory inspection points (MIPS), 182
preflight verification test (PVT), 182
qualification testing, 176 – 178
tables, 177, 178

Radiography test, 179 – 180
Ranger spacecraft, 108
illustration, 109
RDX (herogen; cyclonite), 7
Release devices (*see* Separation/release devices)
Release nuts, 85 – 87
illustration, 86
Retractors, 106 – 108
illustration, 106
Rivers, Vic, 154 – 155
illustrations, 155, 156

Safe-and-arm devices, 129 – 134
illustrations, 131 – 133
Safety standard 208, Federal Motor Vehicle, 150
Sauria, Charles, 145
Secondary explosives, 6 – 7, 27
Separation/release devices:
frangible bolts, 79 – 84
illustrations, 81 – 83
frangible links, 84
illustration, 85
frangible nuts, 77 – 79
illustrations, 77, 80
release nuts, 85 – 87
illustration, 86
Sequential qualification testing, 177 – 178
table, 178
Shaped charges, 41 – 73
conical shaped charge (CSC), 47 – 53
illustrations, 48, 50, 51, 53
table, 47
Shaped charges *(Cont.)*:
flexible linear shaped charge (FLSC), 59 – 73
illustrations, 63, 66, 67, 69, 70, 72
tables, 62, 65
linear shaped charge (LSC), 53 – 59
illustrations, 56, 58 – 61
table, 55
principle of, 41 – 46
illustrations, 42 – 46
Shielded mild detonating cord (SMDC), 34 – 39
illustrations, 35 – 38
SMDC initiator, 16
illustrations, 16 – 17
Sodium azide, 160
Sodium chlorate, 167 – 168
Space shuttle orbiter, jettisonable hatches, 30 – 34
illustrations, 31 – 33
Spit hole, 16, 18, 19, 24
Splice connector, 120, 123, 125 – 127
illustrations, 123, 126
Stab primer, 24 – 25
illustration, 24
Sulfur hexafluoride, 110
Switches, cartridge-activated, 109 – 111
illustration, 110

Testing, (*see* Quality assurance and control)
Tetryl, 7
Thru-bulkhead initiator (TBI), 17
illustration, 17
Thrust-reversing jet, 71
Thrusters, 103 – 106
illustration, 104
Thumping device, lunar, 2
Time-delay compositions, 8 – 9
Torch, welding and brazing, 172 – 174
illustration, 173
Transformer protective switch (TPS), 110 – 111
illustration, 110
Trenching operations, 52
Trinitrotoluene (TNT), 7

Underground distribution splice, 125–127
  illustration, 126
U.S. Department of Transportation, 150

Valves, explosive, 111–113
  normally closed, 112–113
    illustration, 112
  normally open, 111
    illustration, 112

Walker, John, 144
Welding and brazing torch, 172–174
  illustration, 173

XTA (expanding tube assembly), 33–34
  illustration, 33

Zirconium, 136